R0019618072

D1800919

Minerals and Rocks

11

Editor in Chief
P. J. Wyllie, Chicago, Ill.

Editors
W. von Engelhardt, Tübingen · T. Hahn, Aachen

Werner Smykatz-Kloss

Differential Thermal Analysis

Application and Results in Mineralogy

With 82 Figures

Springer-Verlag   New York Heidelberg Berlin 1974

Dr. Werner *Smykatz-Kloss*
Mineralogisches Institut der Universität
7500 Karlsruhe/W. Germany

Volumes 1 to 9 in this series appeared under the title
Minerals, Rocks and Inorganic Materials

Library of Congress Cataloging in Publication Data

Smykatz-Kloss, Werner, 1938—
Differential thermal analysis.
(Minerals and rocks, v. 11)
Bibliography: p.
1. Thermal analysis. 2. Mineralogy, Determinative.
I. Title.
QE369.D5S58   549'.133   74-17490

ISBN 0-387-06906-2 Springer-Verlag New York Heidelberg Berlin
ISBN 3-540-06906-2 Springer-Verlag Berlin Heidelberg New York

This work is subject to copyright. All rights are reserved, whether the whole or part of the material is concerned, specifically those of translations, reprinting, re-use of illustrations, broadcasting, reproduction by photocopying machine or similar means, and storage in data banks.
Under §54 of the German Copyright Law where copies are made for other than private use, a fee is payable to the publisher, the amount of the fee to be determined by agreement with the publisher.
© by Springer-Verlag Berlin · Heidelberg 1974. Printed in Germany.
The use of registered names, trademarks, etc. in this publication does not imply, even in the absence of a specific statement, that such names are exempt from the relevant protective laws and regulations and therefore free for general use. Typesetting, printing and binding: Brühlsche Universitätsdruckerei, Gießen.

Dedicated to
Elvira, Verena, Bettina, and Nadine

Preface

At first glance it may seem presumptuous to want to add yet another to the numerous books on Differential Thermal Analysis (DTA). Thermoanalytical methods have been in use for some time, as shown by the more than five thousand publications containing DTA or TG curves listed by SMOTHERS and CHIANG in the bibliography to their handbook and abstracted in the several volumes of *Thermal Analysis Abstracts* (*TAA*), edited by J.P.REDFERN for the International Confederation for Thermal Analysis (ICTA). Every three years the proceedings of ICTA meetings are published, bringing the latest results of thermoanalytic research. There is also the Scifax DTA Data Index, edited by R.C.MACKENZIE (1962) and modeled on the ASTM pattern card index (used for X-ray investigations), a compilation of the DTA data for several hundred minerals, and inorganic and organic materials. The theoretical foundations of thermogravimetry and DTA have been described in detail by LEHMANN, DAS and PAETSCH (1953), R.C. MACKENZIE (1957, 1970), DUVAL (1963), WENDLANDT (1964), GARN (1965), F.PAULIK et al. (1966), SMOTHERS and CHIANG (1966), and KEATTCH (1969).

Thermoanalytical results are strongly influenced by various factors relative to preparation and equipment (see I-2.4 of this study). This is the reason why we frequently find, in these books as well as in the Scifax-Card catalog, contradictory data on the same substance. Mineralogical publications and textbooks therefore often stress that: "It is possible only in rare cases, and even here only in specific questions, to clearly identify a mineral by means of DTA alone" (GERMAN MÜLLER, 1964); that: "It is not possible to obtain an accurate quantitative determination of the amount of a mineral in a sample by means of the DTA method" (VAN DER MAREL, 1956); or: "Finer distinctions and hence total mineral determination of ceramic raw materials can only be obtained if DTA is supplemented by other analytical methods, specifically X-ray diffraction" (LIPPMANN, 1959). Several authors have published differing DTA curves for similar minerals, and these have sometimes been interpreted to mean that the minerals reflect chemical or genetic differences, even though no such distinctions would be found under truly comparable test conditions. Many publications give incomplete descriptions of sample preparation and analysis and such details are often omitted. A comparison of DTA data on minerals is thus possible only on a limited basis or not at all, so justifying the above-quoted statements of G. MÜLLER, LIPPMANN or VAN DER MAREL. With regard to *mineralogy*, the best textbooks on thermoanalysis frequently merely survey the application of DTA, i.e. they show characteristic curves for particular mineral types and list the often widely differing data found in the literature. It is not often that the causes of these differing data are discussed, as for example they are in the chapter on carbonates by WEBB and KRÜGER in MACKENZIE'S *Differential Thermal Analysis*, Vol. 1 (1970). The disadvantage of all handbooks, even those as carefully compiled as MACKENZIE'S (1957, 1970, 1972), is that they collect and compare the data of many authors obtained under differing conditions of analysis. It is the present author's opinion that, where an analytical method depends so much on the equipment and procedure used as does DTA, *one analyst* should be asked to work out the DTA characteristics of the most important minerals under the conditions recommended by ICTA and other bodies. The assignment should include consideration and standardization of all the factors that can influence the shape of DTA curves and their data. Part II of this monograph tackles this problem by facilitating the identification systematics by means of procedural improvements and supplements (cf. I-2.52; II-10).

Part III of this book concerns special mineralogical applications of DTA beyond the "usual" scope of qualitative and semiquantitative mineral determinations, i.e. the relevance of DTA measurements to the results of crystal chemistry, crystal physics, or petrology. The examples given in Part III for the application of DTA to chemical composition, degree of disorder, or formation of minerals constitute the first steps. There must still be numerous investigative possibilities connected directly with the field of mineralogy. For instance, DTA could be used to some extent for clarification of structural order-disorder phenomena, for detection of high-temperature modifications that are unstable at low temperatures, and for studying variations in thermal behaviour due to mixed-crystal formation, diadochy, etc.

DTA, originally a mineralogic method, is now used more frequently in other fields of scientific research, whereas in mineralogy it is employed almost exclusively for routine determinations. About 1930, more than 80% of all DTA determinations dealt with mineralogical matters; today the proportion of DTA publications relating to all earth sciences amounts to only 2 to 5%. This decrease is clearly evident from the two specialized periodicals, *Journal of Thermal Analysis* and *Thermochimica Acta*. Of the approximately 180 papers given at the *International*

Preface IX

Conference on Thermal Analysis in Davos (ICTA III, August 1971) only 10, and of the 305 papers given in Budapest (ICTA IV, July 1974) only 24 dealt with the geosciences. The emphasis of thermoanalytical research seems to have shifted to the fields of chemistry and materials science. The main reason why mineralogists neglect the DTA method is, in the author's opinion, because the possibilities for the application of differential thermal analysis to mineralogy are still far too little known.

The results of this study were obtained during the years 1966 to 1973 when the author was working in the mineralogical institutes of the universities of Göttingen and Karlsruhe. I wish to thank all the members of these institutes for their support, especially C. W. CORRENS (Göttingen), E. ALTHAUS and H. WONDRATSCHEK (Karlsruhe) for many critical discussions, H. U. BAMBAUER (Münster), R. BEISING (Stuttgart), W. ECHLE (Aachen), F.-J. ECKARDT and P. MÜLLER (Hannover), M. GRAMSE, V. KUPČIK, R. KURZE, G. MENSCHEL, K.-H. NITSCH, and R. USDOWSKI (Göttingen), K. KAUTZ (Essen), H. KULKE (Bochum), F. LIPPMANN (Tübingen), H. MEIXNER (Salzburg), R. METZ (Karlsruhe), G. MÜLLER (Heidelberg), K. H. SCHÜLLER (Selb), and H.-J. TOBSCHALL (Mainz) for sample material, S. EBEL, R. EMMERMANN and U. LENNARTZ for chemical analysis, R. BENDER and R. BENJES for typing the manuscript, U. FRANK, B. LESTI, H. SCHNITTKA, and H. SIEGEL (all in Karlsruhe) for drawing the figures.

Karlsruhe, September 1974   W. SMYKATZ-KLOSS

Contents

Part I. Methods 1

1. Thermogravimetry and Differential Thermal Analysis 1

2. Heat Changes and Their Measurement in DTA 2

    2.1 Cause of Heat Changes 2
    2.2 DTA Apparatus 3
    2.3 Characteristics of DTA Curves and DTA Data 4
    2.4 Factors Which Influence DTA Data 5
        2.4.1 Furnace Atmosphere 5
        2.4.2 Sample Arrangement 6
        2.4.3 Thermocouples 6
        2.4.4 Heating Rate 7
        2.4.5 Reference Material 7
        2.4.6 Grain Size and Packing Density 8
        2.4.7 Amount of Sample 8
        2.4.8 Preparative Factors 10
    2.5 The Technique of Measurement and of Standardization . 10
        2.5.1 Recommendations of ICTA for the Publication of Thermoanalytical Data 10
        2.5.2 Standardization and Indirect Characterization by Means of PA-Curve and Standard Temperature . . 11

3. Calibration and Exactness of Measurement 13

    3.1 Calibration 13
    3.2 Exactness and Reproducibility of Measurements 15
    3.3 Improvement of the Exactness of Measurement Using Internal Standards 16
    3.4 Sensibility of Proof 17

4. Quantitative Determinations by DTA 18

   4.1 Difficulties in Quantitative DTA Determinations of
       Minerals 18
   4.2 Determination of Thermodynamic Data 19
       4.2.1 Equilibrium Temperatures 19
       4.2.2 Heat of Reaction, ΔH 20

5. Methods Combined with DTA 21

   5.1 DTA + High-Temperature X-Ray Analysis 21
   5.2 DTA + High-Temperature Microscopy 21
   5.3 High-Pressure DTA 21
   5.4 DTA + Mass Spectrometer 22
   5.5 Other Methods Related to or Combined with DTA . . . 22

Part II. Application of Differential Thermal Analysis to Mineralogy: Identification and Semi-Quantitative Determination of Minerals 24

1. Elements and Chalcogenides 24

   1.1 Elements 24
   1.2 Chalcogenides 25

2. Halogenides and Sulfates 31

   2.1 Halogenides 31
   2.2 Sulfates 33

3. Oxides and Hydroxides 36

   3.1 Oxides 36
   3.2 Hydroxides 37
   3.3 Soils and Iron Ores 40

4. Carbonates and Nitrates 41

   4.1 Carbonates Free of Water and without Other Anions . . 43
   4.2 Carbonates Free of Water with Other Anions 48
   4.3 Hydrated Carbonates without Other Anions 50
   4.4 Hydrated Carbonates with Other Anions 53
   4.5 Nitrates 57

Contents   XIII

5. Borates, Phosphates, and Arsenates 57

    5.1 Borates 57
    5.2 Phosphates and Arsenates 57

6. Ortho-, Ring-, and Chain Silicates 63

7. Sheet Silicates 64

    7.1 Kaolinites 64
    7.2 Pyrophyllite and Talc 67
    7.3 Montmorines (Smectites) and Vermiculites 69
    7.4 Micas 71
    7.5 Chlorites 73
    7.6 Serpentines 78
    7.7 Palygorskite and Sepiolite 79
    7.8 Clay Minerals with Mixed-Layer Structure 81
    7.9 Mixtures of Sedimentary Minerals ("Clays") 84

8. Zeolites 88

9. Allophane, Opal, and Organic Matter of Soils and Sediments 91

10. Development of Identification Diagrams 93

Part III. Special Application of Differential Thermal Analysis in Mineralogy: Statements about Chemical Composition, Degree of Disorder, and Genesis of Minerals 107

1. Influence of the Chemical Composition on the Decomposition Temperatures of Carbonates and Hydroxides 107

    1.1 Substitution of Ca^{++} by Mg^{++} or Pb^{++} in Calcites . . . 108
    1.2 Substitution of Ca^{++} by Sr^{++}, Ba^{++}, and Pb^{++} in Aragonites 109
    1.3 Substitution of Mg^{++} by Fe^{++} and Mn^{++} in Dolomites 111
    1.4 Hydrozincite and Aurichalcite 114
    1.5 The Incorporation of Al^{+++} into the Structure of Goethite 114

2. Influence of the Chemical Composition on the Temperatures of Structural Transformations 115

   2.1 Carbonates 115
   2.2 Cu-Ag Sulfides 115

3. Influence of the Chemical Composition on the Curie-Temperatures of Magnetites 118

4. Contribution to the Classification of Chlorites 122

5. Smectites and Vermiculites: The Distinction between Di- and Tri-Octahedral Minerals and Grain Size Determination . . . 128

6. Determination of the Degree of Disorder in Kaolinites . . . 130

7. The Interdependence of Degree of Disorder, High-Low Inversion, and Temperature of Formation of Low-Temperature Cristobalites 133

8. The Determination of Inversion Temperatures of Quartz Crystals as a Petrologic Tool 139

9. The High-Low Inversion Behaviour of Microcrystalline Quartz Crystals 146

References 159

Subject Index 173

Part I. Methods

1. Thermogravimetry and Differential Thermal Analysis

Since the study by LE CHATELIER (1887), the thermoanalytical methods have been among the standard methods of mineralogy. From reactions which occur in a mineral or other chemical substance during thermal treatment (heating or cooling), the weight and energy changes can be identified and measured very clearly. Weight loss which for example occurs during dehydration or loss of CO_2, SO_3 etc., can be determined with a thermobalance, recorded, and graphed versus temperature or time *(thermogravimetry, TG)*. For a series of compounds containing H_2O, OH, and CO_2, TG provides characteristic features for identification. With the existence of calibration curves it is also possible to make quantitative H_2O or CO_2 determinations, and from this to tie in chemical mineral analysis, although the errors involved are quite extensive. (Compare DUVAL or GARN).

In *Differential Thermal Analysis (DTA)*, temperature differences ΔT relative to a thermally inert material are measured during heating or cooling of a sample. The DTA curve records these differences during reactions in the sample, showing thermal effects as deviations from the zero line. Coordinates of the resulting DTA diagrams are ΔT (ordinate) and sample temperature (abscissa), both given in °C. Accordingly, whether a reaction requires heat (endothermic reaction) or releases heat (exothermic reaction), the curve slopes toward one side or the other of the zero line. Conventionally, an endothermic reaction is graphed sloping down (see Fig. 2).

Simultaneous DTA and TG may also be possible, e.g. with equipment from Mettler (see WIEDEMANN; TETS and WIEDEMANN) or with the derivatograph (see F. PAULIK et al., 1966a). This is quite suitable for serial analyses. However, from time to time the one method should be waived so that the equipment calibrations can in part be checked.

There is little to add to the explanation of the basis and applications of thermogravimetry given by DUVAL, GARN, or KEATTCH. The TG method is of limited use as compared with DTA. The principle is quite simple; and possibilities for special applications, which for DTA are described in Part III, do not exist for TG. Far fewer mineral characteristics can be obtained from TG curves than from DTA curves. This means that the application of TG to mineralogy is not only less complicated but also less meaningful than DTA. Therefore this book will be restricted to the most important method of thermoanalysis, differential thermal analysis. Care will be taken that the *analytical conditions are kept as reproducible and simple as possible, despite extensive standardization.* In this way the methods described can be practical for DTA apparatus without complicated supplemental equipment (as is necessary for analysis in special gas atmospheres).

In the following sections the most important foundations of the method that are necessary for understanding Parts II and III will be explained briefly. For more precise details see the listed handbooks. Only there, where the DTA method has been improved or supplemented by own work (e.g., see 2.5.2), procedural questions were treated more extensively.

2. Heat Changes and Their Measurement in DTA

2.1 Cause of Heat Changes

When the differences in heat capacity and heat conduction between a sample and a simultaneously heated inert substance are ignored, the sample and inert substance should be at the same temperature during heating as long as there is no reaction in the sample. Such is the case with a small sample and an inert reference material that does not differ greatly from the sample chemically (e.g., with silicate samples, burnt kaolinite should be used as the inert substance).

At the free ends of two thermocouples, the welded ends of which are imbedded in the sample and the inert substance, respectively (see Fig. 1), there is initially no potential. The graph of the difference in voltage as a function of time and/or temperature is parallel to the axis. As soon as an endothermic or exothermic reaction begins, the heating of the sample (and the sample holder block) either remains below or rises beyond the furnace temperature. The sample remains colder (through the use of the supplied heat as reaction heat) or becomes warmer (through heat released by an exothermic reaction) than the inert reference material. The potential is measured at the free ends of the thermocouples' differential connections, and the deviation from the zero line on the DTA curve is thus obtained.

Endothermic, heat consuming reactions include:

    Dehydration (releasing adsorbed water and water bound in the structure as H_2O or OH),

structural decomposition,
melting, evaporation and sublimation reactions,
structural transformations (heating),
magnetic transformations (demagnetizing of a ferromagnetic substance).
reduction.
Exothermic, heat releasing reactions include:
oxidation, including combustion,
processes of freezing,
reconstruction of a crystal structure,
structural transformations (cooling).

2.2 DTA Apparatus

The diversity of DTA equipment in use does not permit immediate comparison of results in different laboratories (LEHMANN, DAS and PAETSCH; DEEG a.o.) because DTA curves and data are influenced by many *equipment factors* (see 2.4). These factors for the different kinds of equipment can be evaluated and compared with the results obtained, as has been shown by numerous workers (see esp. ARENS; BARSHAD; BERG et al.; ERIKSSON; GARN; WIEDEMANN et al.). The model described by all publications for DTA equipment (see Fig. 1) is always the same (e.g., SCHULTZE, pp. 101, 124–127; or the other handbooks). The main parts consist of furnace (1), sample holders (2 and 3), gauges for the regulator

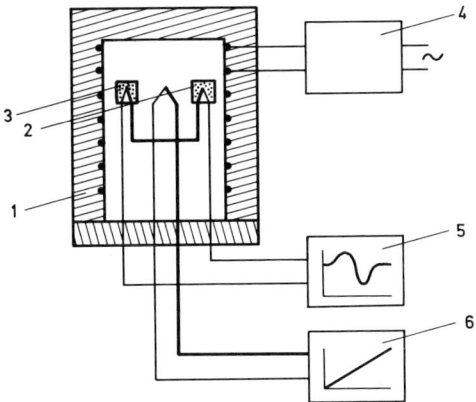

Fig. 1. Model of a DTA apparatus. (After SCHULTZE). Furnace (1) with sample holder crucible [(2) sample, (3) inert substance] and thermocouples; regulator (4), controlled by the furnace thermocouple; recorder (5) and amplifier (6)

(4), amplifier (6), a comparison measuring position and the recording device. The regulator controls the current supply. The thermo-differential potential reaches the recording device through the standard potential amplifier (6).

For the present studies three types of apparatus were used: One, with a nickel block as sample holder, built by LIPPMANN and later modified by means of a photocell compensator and a six-point recorder (built by Hartmann and Braun); a micro-DTA apparatus of Linseis (described, e.g., by KRESTEN); and finally the *Thermoanalyzer-2* by Mettler (see WIEDEMANN). The last two apparatus have platinum-plated sample holders and automatic programm control. All three apparatus are easily heated to 1100° C at a constant rate. Details of the instrumental factors and their influence on DTA characteristics are discussed in 2.4.

2.3 Characteristics of DTA Curves and DTA Data

The temperature differences of a sample when heating or cooling are shown in a DTA diagram as the deviation of the curve from the zero line (base line). An endothermic or exothermic event is apparent as a more or less clearly defined peak. Every such event can be described by the following ten direct DTA characteristics:
1. initial temperature (point A in Fig. 2),
2. amplified initial temperature (point B),
3. peak temperature (C),
4. final temperature (D),
5. temperature difference ΔT (line CE),
6. reaction temperature range (final temperature minus initial temperature, line AD),
7. peak width at half height (reaction width at $\Delta T/2$, line FG),
8. reaction area (shaded),
9. base line shift γ,
10. form (shape, sometimes described as "character"): width, steepness, shoulder, etc.

DTA is a dynamic method. Because the sample is heated continuously the true characteristic reaction temperature (point A), which lies near the thermodynamic equilibrium temperature, is delayed. If the sample were held constant at this initial temperature, e.g. supplying the exact amount of heat necessary for the reaction's completion, then the reaction would run to total completion and no "DTA effect" would be seen. Through "overheating" in DTA the rate of heat absorption through the reaction at point C (therefore at the peak) becomes the same as the differential heat conductivity in the sample, and the reaction is com-

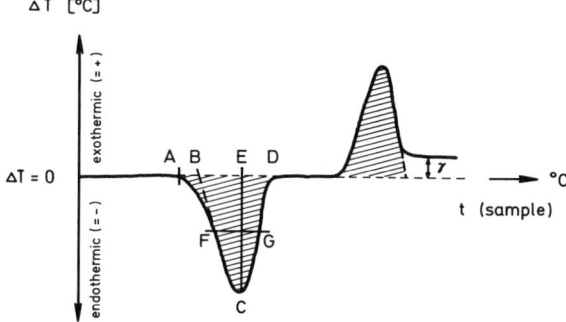

Fig. 2. Characteristics of a DTA curve (explanation in text)

pleted approximately at point G. At D the sample should again have the same temperature as the inert material ($\Delta T = 0$). If this does not occur, as in the case of the exothermic reaction shown in Fig. 2, a base line shift (usually small) has been produced by the difference between the specific heat of the original sample and its reaction products. While the true characteristic reaction temperature, the initial temperature, can be determined only approximately, the *peak* temperature can be quite accurately measured. Although the temperature of the peak depends on several instrumental factors (e.g. different heating rates), it is regarded as the main DTA characteristic. From the area under a DTA peak the reaction heat or heat change, ΔH, can be computed after careful calibration (see 4.2), since it is proportional to the peak area.

2.4 Factors Which Influence DTA Data

2.4.1 Furnace Atmosphere

In order to make the reproducibility of the measurements possible, the analyses of the present study have been carried out with ambient air as the furnace atmosphere. The DTA data therefore reflect this atmosphere. Different DTA characteristics are produced by other furnace atmospheres (CO_2, N_2, Argon etc.). A circulating atmosphere can cause a peak shift. For example, the temperature of decomposition of $CaCO_3$ to $CaO + CO_2$ can be lowered by the elimination of CO_2. Using *ambient (static)* air as the furnace atmosphere the DTA data can be affected by:

a) chemical reactions that occur between air and the sample, its gaseous decomposition products, or the sample holder itself (e.g. oxidation or contamination of Pt containers with sulfides),

b) changes in the reaction mechanism of the sample caused by accessory gases (e.g. peak shifts of dehydration reactions, such as that of gypsum, by differences in the partial pressure of H_2O in the air).

2.4.2 Sample Arrangement

Nickel blocks, platinum containers or crucibles and ceramic blocks are the most frequently used sample holders. According to ARENS, BOERSMA, GARN, LIPPMANN, and WEBB the area under DTA peaks decreases in proportion to the increase in the heat conductivity of the sample holder material. The use of ceramic sample holders therefore results in especially clear DTA peaks and especially large time-lags in the peaks, i.e. higher peak temperatures than with nickel or platinum sample holders.

When using thermocouples imbedded in the sample and thin-walled sample holders (Pt containers), the course of the DTA curve will be influenced primarily by the heat conductivity of the sample. The thermal effects do not increase with this arrangement, but they fade away slower than in the block arrangement. According to LIPPMANN, therefore, platinum containers are preferable for small, intensive thermal effects; but nickel blocks should be used as sample holders in the case of overlapping peaks.

From the work of BOERSMA and LEHMANN et al. it follows that spherical sample holders yield more distinct DTA effects than cylindrical holders. A sample container or the hole in a sample holder block should therefore have a length to width ratio of approximately one, i.e. spheroidal.

Frequently a *drift of the base line* is observed in DTA curves. There are two causes of such a drift: differences in the heat conductivity and heat capacity of the sample and inert substance or a non-isotropic temperature distribution caused, for example, by imperfect centering of the sample holder in the furnace. The second cause is easy to eliminate. The first can, at least in part, be remedied by mixing the sample with inert material or reducing the sample amount.

The development of a "blowout" of a sample part, such as can happen in reactions with intense gas productions (e.g. carbonate decomposition), can be prevented by enclosing the sample in some inert material.

2.4.3 Thermocouples

In most DTA apparatus thermocouples out of Pt-Pt/Rh or Ni-Ni/Cr with a wire diameter between 0.1 and 0.3 mm are used. This size is a compromise between the demand for thin wire in order to decrease the heat conductance and the demand for thick wire to extend the durability of the thermocouples (BOERSMA). Thermal effects are increased when the heat that is absorbed by the thermocouples (e.g., the heat capacity of the thermocouples) decreases. Consequently the soldering pearls of the thermocouples should be as small as possible (LEHMANN).

The *arrangement* of the thermocouples relative to the sample also influences the recorded DTA curves: the most favourable position of the thermocouples will be in the center of the sample and of the inert material. The measurement of $\Delta T = f$ (temp. of sample) permits for instance the exact determination of transformation temperatures independent of the heating rate.

2.4.4 Heating Rate

The heating rate $\Phi = dT/dt$, that means the ascent of temperature in a time unit, influences shape as well as temperatures and ΔT of the DTA peaks. If Φ increases, the peaks will become larger: ΔT is almost proportional to the amount of substance reacting in the time unit. The peaks will get broader because at higher Φ the spacious gradient of temperature in the sample will get larger. At an accelerated heating rate the reactions in the sample need less time. Since ΔT will become larger the peak area $F = \Delta T dt$ will be kept nearly constant. According to SPEIL et al. and KISSINGER the peak temperatures of chemical reactions shift to higher temperatures with increasing heating rates (see 4.2).

DTA of high resolution of which the peak temperatures lie close to the thermodynamic equilibrium temperatures (compare to 2.3) can only be attained at very slow heating velocities. At very rapid heating rates ($>50°$/min) the insufficient resolution of peaks lying close together will only allow coarse determinations. As a matter of experience, for mineralogical investigations heating rates between 5° and 15°/min are suitable.

2.4.5 Reference Material

The reference material for the measurement of temperature differences can be *inert* or *active*. Inert substances have no heat changes in the temperature range $<1100°$ C, its heat conductivity and its specific heat should be similar to the heat properties of the sample. Suitable reference material for silicates, especially for clay minerals, are Al_2O_3 or burnt kaolinite. For the purpose of calibration it is possible to use *active* reference materials: substances with defined transformations at well-known temperatures (simultaneous calibration, see 3.1). Considerable differences in the volumina or in the chemical composition of the sample and the reference material will cause a drifting of the DTA curve away from the zero line (see 2.4.2).

2.4.6 Grain Size and Packing Density

The decomposition temperatures of clay minerals containing OH^- and of carbonates are only dependent on the grain size of the sample if the grains are smaller than $1\,\mu\,\varnothing$: BAYLISS observed decreasing peak temperatures with decreasing grain size, but only in the finest fractions. The heat conductance of a substance depends on its grain size distribution: the high heat conductivity of samples which are packed more densely decreases the temperature gradient in the sample; as a result there is a quicker adjustment of the temperature between sample and sample holder. That means: the DTA curve will be returned to the zero line more rapidly, and the resulting peak will be sharper than in a sample with a smaller packing density. This influence of the packing density has only been observed at low temperatures ($< 600°$ C), in the temperature range of heat conductance. At higher temperatures heat will be mainly transformed by radiation.

2.4.7 Amount of Sample

The amount of the sample used in DTA lies between 0.3 mg (e.g., WITTELS) and some hundred grams (GRIM and JOHNS put whole clods of rocks into the apparatus in order to find a little sulfur or organic matter). In a heated sample there is a large temperature gradient from the wall of the crucible towards the center of the sample. Therefore thermal transformations at a specific temperature do not take place in the whole volumina of the sample exactly at the same moment, but only along a slowly proceeding front of temperature (SCHULTZE).

The amount of the sample determines how long heat will need to cross the whole sample, and as a result there is a kind of "smear-over" of thermal effects over a certain range of temperature. If the sample is too large, a remarkable difference between the temperature of the sample holder and that of the sample will occur.

The temperatures of structural transformations can be determined exactly if the crystals have a low degree of disorder and if the amount of sample material does not exceed 200 mg. But in order to get clear thermal effects the amount of sample must have the necessary size. When extremely small amounts of sample are heated there will be great danger of the heated substance not being representative for the whole rock sample. According to investigations of BERG and RASSONSKAJA and according to the author's DTA runs, the optimal amount of sample material is between 30 and 100 mg; a smaller amount than 20 mg will often give thermal effects too indistinct for clear interpretations.

Measurements of structural transformations make the use of different amounts of sample material unproblematic.

When a *gas* arises as a product of reaction in a heated sample, the progress of the reaction will be controlled by the *partial pressure* of this gas. That is important in dehydration reactions and for the decomposition of carbonates (e.g. see SMITH et al.). Especially the fact that the CO_2 is the reaction product of carbonate decomposition hinders the reaction: the CO_2 will be enriched directly above the sample (at the bottom of the sample holder) because it is much heavier than air, provided that there is no turbulence in the furnace atmosphere. Directly above the sample there is an increased partial pressure of CO_2 which counteracts the progress of the decomposition reaction. The greater the amount of CO_2 arising at such a decomposition reaction (that means: the greater the relation CO_2: metal oxide has been in the heated sample), and the smaller the space the furnace atmosphere can occupy (the less diluted the arising CO_2 is), the more the progress of the decomposition reaction will be retarded. The temperatures of decomposition and dehydration peaks in DTA depend on P_{CO_2} and P_{H_2O} in an exponential manner. If this dependence is not taken into consideration, the peak temperature of decomposition reactions for the same carbonate mineral will vary up to 300° C for different heating conditions. Likewise clay minerals may differ up to 200° C in their dehydration temperatures. This influence is very great, especially when there are small amounts of carbonate sample material. Proof of this is the dependence of the peak temperature of the strontianite ($SrCO_3$) decomposition on the amount of sample material: 2 mg strontianite decompose at 940° C, 5 mg at 990° C, 20 mg at $\sim 1060°$ C and 50 mg at $\sim 1100°$ C, provided that the heating conditions are standardized (compare with 2.5). This peak temperature dependence on the amount of sample, caused by the influence of the decomposition and dehydration temperatures by the partial pressure of CO_2 and H_2O, can be taken for a simple semi-quantitative determination of carbonates and clay minerals out of unknown rock samples (see 2.5.2: PA-curves).

This relatively coarse method has been described almost simultaneously by both WATERS and SMYKATZ-KLOSS (1967a). Later GROSS used it to estimate small amounts of spurrite besides large amounts of calcite from DTA curves of rocks containing both minerals. It is above all a method suitable for carbonate minerals being decomposed at very high temperatures ($>1100°$ C): so the minerals strontianite, witherite, norsethite, alstonite or benstonite, which cannot be investigated in a "normal" DTA apparatus (heating possible up to 1000° C), can be determined by using very small amounts of sample (3–15 mg; see SMYKATZ-KLOSS, 1967b).

2.4.8 Preparative Factors

Besides the apparative factors (e.g. heating rate, furnace atmosphere, thermocouples etc.) and besides the properties of the substance to be analyzed (packing density, grain size), the *preparation* of the sample may have some influence on the appearance of the DTA curve. The crystal structure of quartz will for instance be destroyed when ground too long in an agate mill (HOFMANN and ROTHE). That means: by increasing time and intensity of grinding the defect character of the crystals will increase, too, so changing the physical properties of the crystals (e.g. the high-low inversion temperature of quartz crystals, see III-8). The dehydration behaviour of clay minerals is influenced by the humidity of the air of the laboratory and so on. In order to get comparable and reproducible DTA results the influence of preparative factors must always be considered.

2.5 The Technique of Measurement and of Standardization

2.5.1 Recommendations of ICTA for the Publication of Thermoanalytical Data

The ignorance of all the factors influencing DTA curves and discussed above is the reason that you will find contradictory DTA data for a lot of minerals in many mineralogical publications. And these contradictory results have brought into discredit all DTA, even among mineralogists. This is the reason for the appeals of the *International Confederation for Thermal Analysis* (ICTA) to note all conditions of analysis in publications of thermal analytical data completely and carefully. Without knowledge of the analytical and preparative conditions in use, the DTA curve of a mineral will be nothing more than a vague survey of its thermal behaviour not comparable with other DTA results published. In thermal analysis the influence of all the apparative and preparative factors is much greater than in other mineralogical techniques (e.g. the X-ray methods). For this reason the reference to the recommendations of the ICTA (see McADIE, WIEDEMANN, HEIDE) has to be emphasized in the following text. In the present monograph these recommendations will be followed. DTA publications should contain the following information:

1. Origin and description of the apparatus used, including the geometry and composition of the thermocouples.

2. Information about the size, the geometry and the material of the crucible and of the crucible holder.

3. Characterization of all used samples, reference and dilution material by a clear mineralogical name or chemical designation (formula).

4. Information about the origin, the treatment and the purity of these substances (if known).

5. Amount of sample and amount of dilution material.

6. Average values of the linear heating or cooling rate during the temperature range where the studied reactions in the sample will take place.

7. Description of the gas atmosphere, of the atmospheric humidity, of the atmospheric pressure of the samples surrounding.

8. Information about the methods used for the identification of intermediate and final reaction products.

9. Reproduction of the original curves (if possible) including the reference scale.

10. In the case of reproduction of DTA curves: real scale of time or temperature of the abscissa from left to right.

11. On the ordinate: temperature difference (ΔT) in $°$C, endothermic reactions downward and exothermic reactions upwards.

12. Identification of all thermal effects (if possible).

2.5.2 Standardization and Indirect Characterization by Means of PA-Curve and Standard Temperature

There are two consequences resulting from the discussion of all factors influencing the DTA data: firstly each published DTA curve should be completed by the information about all conditions of analysis (see 2.5.1), secondly as much as possible mineralogical DTA investigations should be made under *standard conditions*. Indeed there exists an "ICTA Committee on Standardization", and in some handbooks of thermal analysis (e.g. MACKENZIE or SCHULTZE) some hints are to be found on the necessity of standardized investigations in order to get comparable and reproducible results. But the progress of standardization techniques is very slow, and according to SCHULTZE in a lot of cases standardization would hinder the attainment of optimal results. Absolutely necessary in the case of investigating a series of similar samples is always to analyze under the same conditions (e.g. in the study of the inversion behaviour of different quartz crystals). The DTA apparatus fabricated and used in different countries are of course somewhat different, and in some cases many problems may arise on the way to standardize the DTA techniques. But the possible disadvantages in standardizing seem meaningless compared with the possibility of attaining DTA curves and data that have been made worldwide comparable by standardization. In the author's opinion DTA as an identification method in mineralogy would win enormously, because then it would no longer be possible to determine—for instance— the decomposition temperature of calcite by means of DTA both at 600° and 950°C. Therefore a scheme of standardization conditions will be proposed in the following text, although this proposal cannot be more then the first step to making completely comparable DTA results, even for quite different DTA apparatus, possible. These proposed standard conditions (Table 1) were the best possible conditions for the normal thermal analysis of minerals in more then a thousand DTA runs. For special investigations in mineralogy the standard conditions have to be modified (see part III).

Table 1. Proposed standard conditions of the DTA of minerals; treatment of *very hard samples* ($>6^1/_2$ after the Mohs sample material scale): 3 min grinding by hand in a corundum mortar (2 cm$\emptyset$); *hard samples* ($4^1/_2$–$6^1/_2$): 1 min grinding in an agate mill (machine) at low velocity (mill: 500 g volume, 3 agate balls of <2.5 cm $\emptyset$); *soft samples* ($<4^1/_2$): 2 min pulverizing by hand in an agate mortar (~ 10 cm $\emptyset$)

Amount of sample material	100 mg + 20 mg Al_2O_3 (mixed)
Reference material	150 mg annealed Al_2O_3
Grain size of sample material	Clay minerals: 0.6–2 μ $\emptyset$ Other material: 60–200 μ $\emptyset$
Situation of sample material	Thermocouples exactly in the center of the sample, with direct contact to the heated substance
Furnace atmosphere	Air, without any current or turbulence
Sample holder	Pt crucible $\emptyset \sim 5$ mm; relation length: diameter <4; wall thickness 0.1 mm
Thermocouples	Pt-Pt_{90}/Rh_{10}; $\emptyset$: 0.1–0.3 mm
Heating rate	10°/min after 300°C
Packing density	Loose packed, no pressing
Additional conditions as mentioned in 2.5.1	

In monomineralic samples the amount of 100 mg is to be prepared in a simple way. But most of the mineralogical substances consist of mixtures of minerals ("rocks") in which the one mineral which can be determined by DTA might only occur in a few percent. In dehydration and decomposition reactions of which the peak temperatures depend strongly on the amount of sample material (see 2.4.7), this dependence makes it possible to determine the portion of the sample that is dehydrated or decomposed during the heating process by using a new (indirect) DTA characteristic. This is the *curve of sample amount dependence* (= PA-curve, from the German „*P*roben-*A*bhängigkeit", compare SMYKATZ-KLOSS, 1967). Such PA-curves must be available from calibrating dehydration and decomposition peaks of each mineral: because of different H_2O (resp. OH) or CO_2 contents of the minerals, the gradient of the diverse PA-curves will differ from each other, so reflecting the different peak temperature dependence on the amount of a mineral in a heated sample. From these PA-curves it is possible to determine the amounts of minerals in an unknown sample if you succeed in finding the (known) PA-curve belonging to the mineral being dehydrated or decomposed in this unknown sample. In each case standardized heating conditions are necessary (Table 1). A lot of such PA-curves of clay minerals, carbonates, phosphates and some other minerals can be found in part II of the present monograph: the simple logarithmic diagrams contain the curves of peak temperature dependence on the amount of sample of these minerals for 10, 50, and 100 mg, and these PA-curves can be used as calibration and determination curves. The fault of this method is $\pm 5\%$ of the determined amount for small amounts of minerals to be determined (that means in samples consisting of fewer than 20% of this mineral). For amounts of > 50 mg (= 50% in a standard sample amount of 100 mg) the fault increases to $\pm 10\%$ of the determined amount (for detailed discussion see SMYKATZ-KLOSS, 1967 a, b).

The extrapolation of these PA-curves to low amounts of sample material leads to another indirect DTA characteristic, to the *standard decomposition (resp. dehydration) temperature*, which represents an amount of sample of 1 mg (that is the beginning of the logarithmic scale on the abscissa, compare the figures in part II). These standard temperatures should be very close to the thermodynamic equilibrium temperatures (compare 2.3). For the DTA of minerals, therefore, besides the measurement of the direct characteristics peak temperature, peak area, ΔT and shape of a peak, the construction of the indirect DTA characteristics PA-curve and standard temperature can be helpful, especially in mixtures of different minerals like in soils and some sediments.

3. Calibration and Exactness of Measurement

3.1 Calibration

Each DTA apparatus should be calibrated before starting with investigations and later after almost 200 runs. For caloric measurements (see 4.2), a still more frequent calibration is necessary by heating of materials with well-known transformation temperatures and heats of reaction. Table 2 contains some substances and their transformation data suitable for calibration of a DTA apparatus following the intentions of BARSHAD, D'ANS-LAX, WIEDEMANN and VAN TETS, and SMYKATZ-KLOSS (1967b). The determination of *melting points* (e.g., those of metals, see WIEDEMANN and VAN TETS) has to be made in graphite crucibles with thermocouples lying directly below the crucibles.

Table 2. Calibration materials for DTA

Substance	Kind of transformation	Temperature (°C)	Heat of reaction (cal/g)	ΔH-fusion (kcal/Mole)
Gallium	Melting point	29.8		1.335
NH_4NO_3	Structural transf.	32		
NH_4NO_3	Structural transf.	85		
NH_4NO_3	Structural transf.	125		
KNO_3	Structural transf.	127.8		
Na_2SO_4	Structural transf.	147		
AgJ	Structural transf.	147		
$AgNO_3$	Structural transf.	160		
NH_4NO_3	Melting point	170		
$AgNO_3$	Melting point	212	16,7	
Na_2SO_4	Structural transf.	215		
Tin	Melting point	213.9 + 0.001		1.68
Bismuth	Melting point	271.0		2.63
AgCl	Melting point	307		
$NaNO_3$	Melting point	314	45,3	
Lead	Melting point	327.4		1.141
KNO_3	Melting point	339	25.3	
Ag_2SO_4	Structural transf.	432		
AgJ	Melting point	552		
Na_3AlF_6 (cryolite)	Structural transf.	562.7		
K_2SO_4	Structural transf.	583.5		
Na_2MoO_4	Structural transf.	642		
Ag_2SO_4	Melting point	652		
Aluminium	Melting point	660.0 ± 0.2		2.57
Na_2MoO_4	Melting point	687		
KCl	Melting point	775		
NaCl	Melting point	801		
Witherite, $BaCO_3$	Structural transf.	810 ± 1.0		
Silver	Melting point	961	25.0	2.70
Norsethite, $BaMg(CO_3)_2$	Structural transf.	968		
Witherite, $BaCO_3$	Structural transf.	980		

For the *simultaneous calibration* in which the substance of calibration is a part of the heated sample, all substances will be suitable of which the thermal effects lie close to DTA effects of the sample minerals. It is possible, too, to mix the calibration substance with the reference material. This will be of advantage if the materials of calibration and investigation show both endothermic or exothermic reactions of comparable temperature and intensity. By measuring the thermal effect of the reference material against that of the sample material, the DTA curve only marks the difference of both effects; by comparing the area of this difference peak with the area of the pure calibration material peak, the heat of reaction of the heated sample mineral is to be determined (see 4.2).

3.2 Exactness and Reproducibility of Measurements

In investigations of the high-low inversion temperatures of quartz crystals some analysts attained a reproducibility of $\pm 0.1°$ C of these inversion temperatures. That means: in several DTA runs of the same substance the deviation of the determined temperature has been not more than 0.1° C (KEITH and TUTTLE; SMYKATZ-KLOSS, 1970; KRESTEN, 1971b). Indeed, special arrangements exist for thermal analysis (e.g., in the apparatus of Hartmann and Braun or Mettler) which allow small voltages to be measured with an accuracy better than 0.2° C, but for "normal" DTA apparatus such a reproducibility seems to be very good.

For one of the three apparatus used in the present study the *exactness of measurement* can be followed up on account of investigations of WEISSE, HUNSINGER and STÄRK, for the self-built apparatus of LIPPMANN, completed by a photocell compensator from Hartmann and Braun.

Thermocouples of Pt-Pt/Rh of course have the disadvantage that their thermoelectric power is small compared with that of other thermocouples, but these Pt-Pt/Rh-thermocouples exhibit the best steadiness of all used thermocouples. Moreover, they are nearly insensitive to ageing (WEISSE). The lower thermoelectric power can be intensified by a suitable amplifier combined with an automatic photocell compensator and a high-sensitive recorder which note the thermal effects in a temperature-time scale.

The photocell compensator has the characteristic that each range of measurement begins with 0 mV adequate to the current 0 of the assistant ammeter. By measuring a temperature of 1000° C (being represented by nearly 10 mV with Pt-Pt/Rh thermocouples), the range of measurement has to reach from 0 to 10 mV. If the ammeter has a margin of error of $\pm 0.5\%$, this temperature of 1000° C can be measured with an accuracy of $\pm 0.5\%$ or $\pm 5°$ C. By putting a resisting force into the circuit of the thermocouples (see HUNSINGER) you will be able to obtain a *"suppression of the zero-point"* in the photocell compensator, that means the ammeter will only reach from 9 to 10 mV over the scale and no longer from 0 to 10 mV, so reducing the error to 1/10 of the first temperature range (0–1000° C): 0.5% of 100° C (= 9–10 mV) instead of 1000° C, means a mistake of only $\pm 0.5°$ C. By combining the photocell compensator with a point recorder and this suppression of the zero-point (as is shown in two of the three apparatus used in the present study), the *exactness of $\pm 0.5°$ C* in the temperature range up to 1000° C can be reached for thermocouples of Pt-Pt/Rh. If the apparatus has been individually calibrated by structure transformation temperatures or melting points, the margin of error can be reduced to $< \pm 0.3°$ C (HUNSINGER). A detailed estimation of all resources and quantities of error appearing in such an apparatus is made by WEISSE:

1. Error f_1 of the equipment of compensation = $\pm 0.1°$ C;
2. error f_2 at the recording ammeter = $\pm 0.25°$ C;
3. f_3, resulting by switching on the counter voltage = $\pm 0.03°$ C;
4. f_4, by variation of the light spot galvanometer = $\pm 0.04°$ C.

In the case of unfavourable conditions the errors f_1, f_2, f_3, and f_4 may sum up, resulting in a total error of $\pm 0.42°$ C in the temperature determination (at a supposed temperature of 700° C). This total error can be reduced to $F = \pm 0.27°$ C by installing a supplementary precision milliammeter and by using a most carefully manufactured point recorder (WEISSE). This exactness of $\pm 0.27°$ C is valid for normal temperature determinations by means of a calibrated DTA apparatus equipped with a photocell compensator with the possibility of suppression of zero-point and with a multipoint recorder. The apparatus used made by Hartmann and Braun or up-to-date apparatus from Mettler or Stanton will attain this exactness.

3.3 Improvement of the Exactness of Measurement Using Internal Standards

In special investigations the use of *internal standards* still permit a somewhat higher exactness of measurement, as discussed in 3.2. An internal standard is a substance with very wellknown thermal properties (peak temperatures or heats of reaction), and this substance has to be added to the heated sample. The peak temperatures of the internal standard material should lie close to those of the sample minerals. For the investigation of the high-low inversion behaviour of Cu_2S and Ag_2S which invert between 70 and 105° C resp. at nearly 180° C (compare with

III-2.3), quite a lot of well-known substances used as internal standards allow the construction of a very exact calibration curve mV towards peak temperature (e.g., NH_4NO_3, KNO_3, Na_2SO_4, AgJ, $AgNO_3$, AgCl, see Table 2), and by using it the inversion temperatures of both sulfides can be determined with an accuracy of $\pm 0.2°$ C.

In the case of the investigation of the high-low inversion behaviour of natural quartz crystals, Na_3AlF_6 (the mineral cryolite) and K_2SO_4 can be used as internal standards. Both substances show inversion temperatures, too, lying almost 20° C remote from each other (562.7 resp. 583.5° C on heating) and including the inversion temperature of a "normal" quartz crystal ($\sim 573°$ C). In the temperature range between 550 and 580° C the inversion temperature of quartz crystals can be determined with an error margin of $\pm 0.2°$ C using one of the two substances as internal standard, usually K_2SO_4, giving the most clear peak (compare with III-8). Necessary for such exactness in temperature determination by means of DTA with internal standards are the existence of useful standards with well-known thermal effects and sharp, easily measurable peaks both for the standard and for the sample substance.

3.4 Sensibility of Proof

Generally the sensibility of proof in DTA is very much greater for dehydration and decomposition reactions than for structural transformations. While the high-low inversion of quartz is only to be observed in a DTA curve if the sample of 100 mg contains more than 10 mg quartz, the same DTA curve of 100 mg sample amount will still reflect 0.3 mg gypsum, 0.3 mg of a carbonate mineral and 3 mg of a clay mineral, all being characterized by dehydration or decomposition reactions. The combustion of bitumen and other organic matter is so strong that the DTA curve will clearly show 0.1 mg of organic matter. LINSEIS used the DTA to determine amounts of 0.1–0.3 mg gypsum in ceramic raw materials, and similar amounts of sulfur will certainly be discovered by DTA, too. By means of special apparatus (e.g. a microapparatus, see LINSEIS or KRESTEN, 1971 b) or some thermocouples arranged one after another, the minor thermal effects of many structural transformations (even by small amounts of an inverting mineral, for instance 2–10 mg in a sample of 100 mg) can be clearly recognized. Especially for the decomposition reactions of carbonates and some dehydration reactions of phosphates or sulfates, the differential thermal analytic sensibility of proof is so good that the DTA will be superior to most other methods of mineral identification. This statement should not be overestimated: as a rule the identification of a mineral by means of X-ray methods will be better than by means of DTA, but a lot of minerals exist which can be identified better by DTA than by means of X-ray: for instance strongly disordered minerals out of soils and sediments and, naturally, the X-ray amorphous species opal, allophane and limonite.

4. Quantitative Determinations by DTA

4.1 Difficulties in Quantitative DTA Determinations of Minerals

In the quantitative determination of minerals the problem is (after ARENS) to find out a relation between any measurable DTA effect and an adequate quantity of a mineral in the heated sample being responsible for this effect. ORCEL, AGAFONOFF and ORCEL and CAILLÈRE have compared the appearance and the temperature of dehydration reactions of unknown clay samples with those of well-known samples. So they received approximate data about the portions of clay minerals in these unknown samples. NORTON compared the peak heights ($=\Delta T$) of DTA curves of unknown samples with those of calibration mixtures, but because ΔT does not only depend on the *amount* of a reacting substance during the heating process, but also on the chemical composition of the reacting mineral, on the disorder of the structure of this mineral and on some apparative factors (e.g. on the heating rate), this measurement of ΔT is unsuitable for quantitative measurements (LEHMANN, DAS and PATSCH). Meanwhile only the *area* of a deflection in the DTA curve is regarded as a measure for the amount of a reacting substance during the heating process (BERKELHAMER and SPEIL; HENDRICKS and ALEXANDER; KERR and KULP; SPEIL et al.; ROKOSZ, PAULIK et al. and others). The area is proportional to the amount of heat m used or set free by the thermal reaction which is responsible for this deflection (or peak) area:

$$\int \Delta T dt = f(m).$$

The use and interpretation of this relation is, however, not without difficulties (ARENS), since even in DTA curves showing only one deflection it is not simple to construct the base line (zero line) exactly: there are differences in the specific heat between the original sample and the reaction product. These differences are responsible for a shifting of the zero line (see 2.3). The area of a DTA deflection (peak) not only depends on the amount of a reacting material (resp. on the heat used by this reaction), but — so ARENS — on the reaction time (and the reaction time itself depends on the heating rate!), on the heat of reaction and also on the heating rate. For a detailed discussion of this point see ARENS.

The measurement of the areas will be almost impossible if some thermal effects superimpose, producing an overlapping of peak areas in the DTA curve. This is the case, for instance, in most of the natural samples. But the margin of error is great in a lot of monomineralic samples, too, as in some kaolinitic clays or carbonates: an error of < 10 weight-% will only seldom be obtained by this method. For this reason the DTA can only be called a *semi-quantitative* method. The exception are some special cases in which extensive work of calibration improved the quantitative results (e.g. LINSEIS or CARPENTER).

4.2 Determination of Thermodynamic Data

4.2.1 Equilibrium Temperatures

As a rule methods exist that are more reliable than DTA for the determination of thermodynamic data. But in several cases thermodynamic data can be determined by DTA with sufficient accuracy, so far as no variations of pressure appear. Thermodynamic equilibrium temperatures can be measured by the method of ARENS or TANAKA (see SCHULTZE), by varying the heating rate and extrapolating the resulting curves to the heating rate $\Phi = 0$ (see Fig. 3). To achieve this it is necessary to work at very low heating rates.

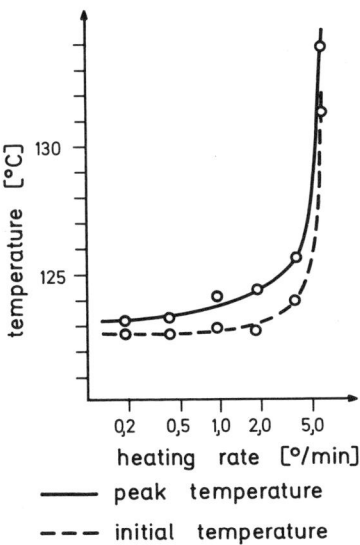

Fig. 3. Determination of equilibrium temperatures by varying the heating rate. (After TANAKA, from SCHULTZE)

4.2.2 Heat of Reaction, ΔH

About the theoretical background of the ΔH-determination see ARENS, SCHULTZE, MACKENZIE, GARN, WENDLANDT or LUGSCHEIDER, After caloric calibration of the DTA apparatus either by simultaneous calibration or by measuring the melting temperatures of some metals with well-known heats of reaction (see WIEDEMANN and VAN TETS), the factor of proportionality

$$K = \frac{\text{supplied additional amount of heat}}{\text{peak area}}$$

has to be determined by electric calibration (compare SCHULTZE, p. 81: measurement of peak areas at different supplies of current). Then ΔH can be calculated after

$$\Delta H = \int_{t_1}^{t_2} K \Delta T \, dt$$

The totally supplied additional amount of heat between t_1 and t_2 (= initial and final temperature of a peak; begin and end of the additional heat current) for running down of the reaction is identical with the heat of reaction. Since the coefficient K of heat transformation can be considered as constant in the relatively narrow temperature range the rundown of the reaction needs, K can be replaced by the average value $\bar{K}$, which has to be determined only once for each apparatus and comparable heating conditions. So becomes

$$\Delta H = \bar{K} \int_{t_1}^{t_2} \Delta T \, dt.$$

Besides this calculation, heats of reaction can also be found out *graphically*, similar to the determination of equilibrium temperatures (SCHULTZE) by measuring the peak areas at different heating rates and extrapolation to a heating rate $\Phi = 0$. Some heats of reaction so determined from a compilation by ARENS are:

	ΔH, cal/mole
Quartz, high-low inversion	180[1]
Goethite, dehydration and decomposition	19000
Kaolinite, dehydration and decomposition	77400
Illite, dehydration ($\sim 550°C$)	24000
Montmorillonite, dehydration and decomposition	31000
Calcite, decomposition	42500
Aragonite $\to$ calcite transformation	390

[1] After FAUST (1948): 3.1 cal/g.

The differential thermal analytic determination of additional thermodynamic or kinetic data is still possible, even if a little more complicated than the determination of equilibrium temperatures (WENDLANDT). But the error in calculating the activation energy or the enthalpy of chemical reactions is mostly so great ($\pm$ 10–25% of the determined values) that the work with DTA can only be regarded as *semi-quantitative*. For a detailed discussion about this aspect, see the handbooks quoted (MACKENZIE, SCHULTZE, WENDLANDT, GARN, McLAUGHLIN).

5. Methods Combined with DTA

5.1 DTA + High-Temperature X-Ray Analysis

Many crystalline substances show at distinct temperatures spontaneous, reversible structural transformations, the existence and temperatures of which can be observed very clearly in a DTA curve. However, the nature of the resulting products of reaction remains obscure, since these transformations receded during cooling. In these cases a combined DTA—X-ray apparatus like that described by WEFERS will be very helpful, in which the continuous heating is interrupted from time to time so that the temperature remains constant as long as the recording of an X-ray diagram is needed. In practise the apparatus consists of an X-ray apparatus with a sample holder suitable for high-temperatures, which serves simultaneously as a DTA furnace combined with the remaining parts of a DTA apparatus. The method of WEFERS makes it possible to identify reaction products with a short existence that are only stable over a small range of temperature.

5.2 DTA + High-Temperature Microscopy

For optical studies of fabric changes appearing after thermal effects during heating of metals, DICHTL and JEGLITSCH combined a DTA apparatus with a high-temperature microscope. In mineralogy, above all thermal optical investigations have been applied to investigations of structural changes of salt minerals (e.g., see HEIDE, 1967).

5.3 High-Pressure DTA

WIEDEMANN (1964) published DTA curves of calcium oxalate heated under pressures of 10^{-5} to 720 Torr; he showed exothermic effects being influenced very strongly by small pressure variations. WEBER and ROY investigated the dehydration behaviour of kaolinites in dependence on pressure in an apparatus suitable for pressures up to 10000 p.s.i. and for temperatures up to 650° C.

Additional DTA work has been done in cold-seal pressure vessels as well as in piston-cylinder apparatus, and a few melting curves have also been measured in high pressure DTA apparatus, e.g., see WYLLIE and TUTTLE or WYLLIE and RAYNOR. Recently TURNER described DTA in a piston-cylinder apparatus up to 6 kbar and 270° C, and ROSENHAUER et al. (1974, in press) reported high-pressure DTA studies on pyroxenes up to 7 kbar and 1400° C. However, all these petrologically very interesting results demonstrated that this combined method still includes some problems, as for instance the convection currents sometimes appearing at temperatures higher than 400° C. For the purpose of this monograph there is no point in going into more detail, as this will be done for a projected study on quantitative DTA concerning petrological problems and including high pressure measurements.

5.4 DTA + Mass Spectrometer

By combining a mass spectrometer with a heating appliance LANGER and GOHLKE obtained the new investigation method MTA (= mass spectrometric thermal analysis). By adding a DTA apparatus the MTA was developed into the MDTA (mass spectrometric differential thermal analysis, see LANGER, GOHLKE and SMITH). This method enables the immediate determination of volatile reaction products: a small part of the gas which has formed on heating in the DTA furnace enters the ionization chamber of the mass spectrometer through a narrow capillary and a valve, where it can be investigated immediately. A recorder shows the thermal effects as well as the composition of the gaseous reaction products in dependence on the heating temperature. The method could be applied to geochemical investigations of gaseous and fluid inclusions in minerals. For this purpose an apparatus constructed by MURPHY, HILL and SCHACHER seems to be the best one, as it allows the analysis of whole the gaseous products forming during the heating of a sample.

5.5 Other Methods Related to or Combined with DTA

For special chemical analysis a lot of other methods exist combined with DTA (e.g., see WENDLANDT). Very suitable for mineralogical problems is the *derivatograph* constructed by F. PAULIK, J. PAULIK, and

L. ERDEY (1966a), an apparatus for simultaneous DTA, TG and DTG (= differential thermo-gravimetry). Another method, the *isothermal differential thermal analysis* described by COHEN, KLEMENT and KENNEDY, can be used for the study of phase transitions caused by changes of pressure or partial pressure of a component at constant temperature, and for the study of melting point determinations under high pressures. In this method samples were heated to a certain temperature, which will be kept constant while pressure is changing.

The sensibility and the selectivity of DTA can be increased if, in addition to the DTA curve, its first derivation after the time is interpreted:

$$d\Delta T/dt = f(T).$$

This happens automatically in the *derivative differential thermal analysis* (*DDTA*, see F. PAULIK, J. PAULIK and L. ERDEY, 1966a–c). The method is very suitable for the study of weak thermal effects, as are some structural transformations (GARN). The *differential calorimetry* permits the direct measurement of calorimetric data (reaction enthalpies, specific heats etc.). For physico-chemical investigations a lot of partly complicated apparatus related to DTA exist, for instance one constructed by WATSON et al. (and built by Perkin-Elmer) for "*differential enthalpy analysis*". The methods mentioned up to now are to be applied to mineralogical problems. The following methods related to DTA are frequently applied in chemical and physico-chemical investigations. Besides the combination of DTA with a mass spectrometer (5.4), other combined methods of DTA-gas analysis are gaschromatography/DTA, gas density/DTA, caloric conductivity/DTA (GARN), thermomanometry and thermal gasvolumetry (BERG, 1961). Finally, F. PAULIK, J. PAULIK and L. ERDEY (1966b) have developed an arrangement to be added to the derivatograph which allows the simultaneous investigation of differential thermal and dilatometric behavior of a sample.

Part II. Application of Differential Thermal Analysis to Mineralogy: Identification and Semi-Quantitative Determination of Minerals

1. Elements and Chalcogenides

1.1 Elements

Among the minerals occurring in elementary form, only sulfur (S) and graphite (C) are to be determined by DTA. The metals silver, gold, platinum, copper or bismuth can be recognized differential thermal analytically at their temperatures of melting (e.g., Ag melts at 960° C), but for one thing the identification of such expensive minerals by means of melting temperature determinations in DTA would be paradoxical, secondly these metals occur very seldom in chemical pure substances. *Sulfur* shows a weak endothermic reaction at 130° C (melting point, after KOPP and KERR, 1957), and a very strong exothermic deflection with a peak at 380° C (oxidation of S to SO_2). The author's DTA runs of greater sensibility show the endothermic peak caused by melting at 120° C (melting point, after D'ANS-LAX: 119° C), and at 96° C another weak endothermic peak occurs reflecting the structural transformation from orthorhombic to monocline. The oxidation peak, occurring at 380° C for *50* mg in our runs, is the most important DTA characteristic for elementary sulfur as well as for sulfides, but in the last case it appears without exception at higher temperatures (see 1.2). This very intensive oxidation deflection (peak) allows very small quantities of sulfur in a rock sample to be determined. GRIM and JOHNS heated some hundreds of grams to find small traces of sulfur. It would be more simple to heat the amount of 150 mg, finely ground, consisting of 50 mg sulfur containing rock sample mixed with 100 mg of inert material (compare with 1.2).

The combustion of fine-grained *graphite* causes an extremely strong exothermic deflection (see Fig. 4): between 400 and 600° C with a peak temperature which can rarely be determined exactly. Small amounts of graphite can be recognized in the DTA curve even better than small amounts of sulfur or sulfides; the exothermic peak of graphite can be

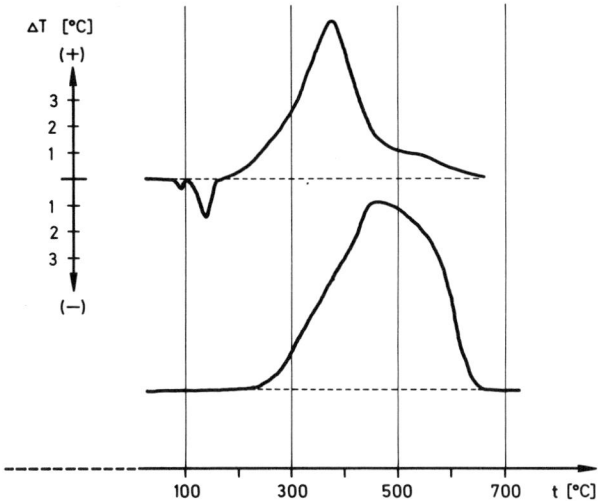

Fig. 4. DTA curves of 50 mg sulfur + 100 mg Al_2O_3 (above) and of 50 mg graphite + 50 mg Al_2O_3 (below). Sulfur from Agrigento, Sicily

distinguished from that of sulfur or sulfides by its shape (form) in the DTA curve: while the oxidation of S and chalcogenides generally causes relatively sharp deflections (see Fig. 4–7), the oxidation peak of graphite or of organic matter (see II-9) is broad and rounded, without any clearly perceptible top (peak). The distinction by DTA between graphite and organic matter will of course not be very simple (see II-9). Conditions of analysis see 1.2.

1.2 Chalcogenides

In DTA of sulfur and chalcogenides (= sulfides, selenides, tellurides, arsenides, antimonides), some precautions must be arranged because the decomposition of minerals sets free gaseous reaction products (S, SO_2, SeO_2, etc.) which are generally very aggressive, as they tend to combine with metals very quickly. So the endothermic peak at 680° C, observed in DTA runs of pyrite (KOPP and KERR, 1958; SMYKATZ-KLOSS, 1966), shows that the surface of the nickel sample holder used in these runs has been partly altered to Ni_3S_2 (heazlewoodite, COLE and CROOK). And crucibles or thermocouples of platinum are quickly poisoned by alloying with sulfur (PtS, cooperite formation!) or other volatiles. It is possible to protect crucibles, thermocouples and sample holders from alloying by coating them with thin cylinders of Al_2O_3 or ceramic material (HILLER and PROBSTHAIN, 1955; KOPP and KERR, 1957; DUNNE and KERR, 1960),

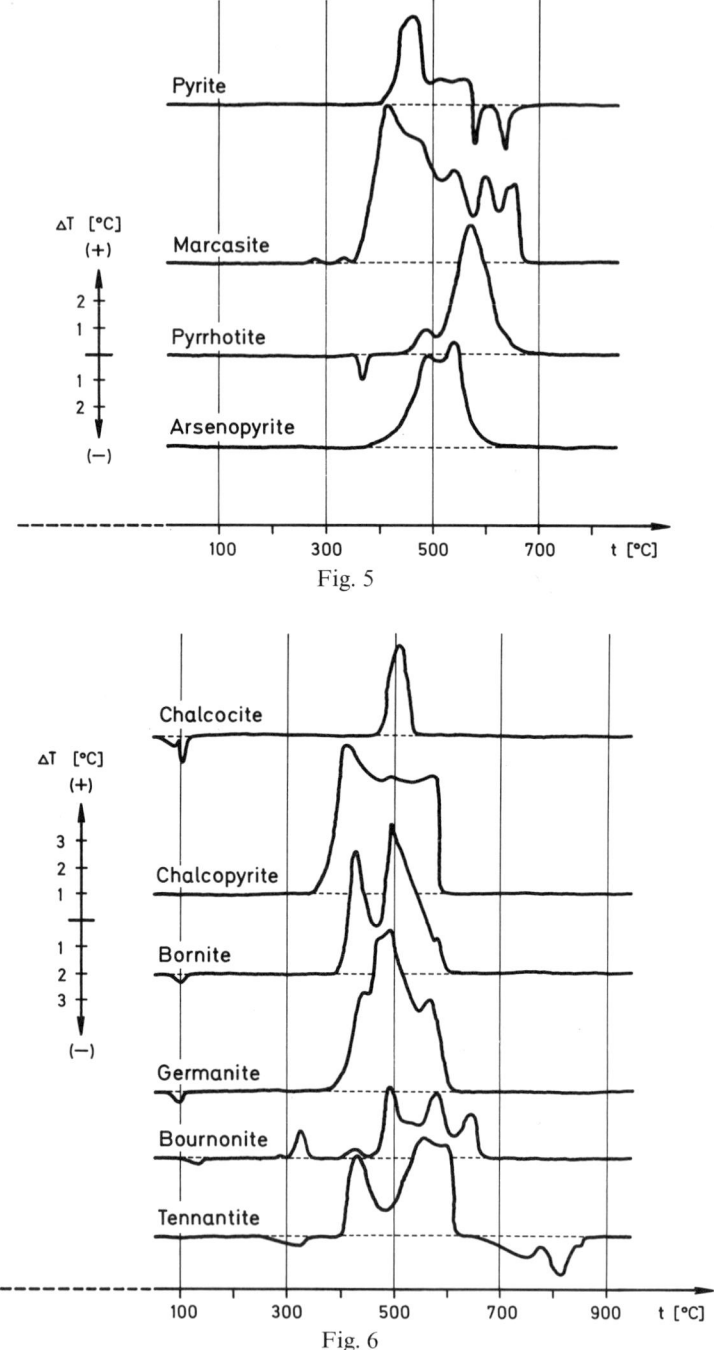

Fig. 5–7. DTA curves of Fe- (Fig. 5), Cu- (Fig. 6) and other sulfides (Fig. 7)

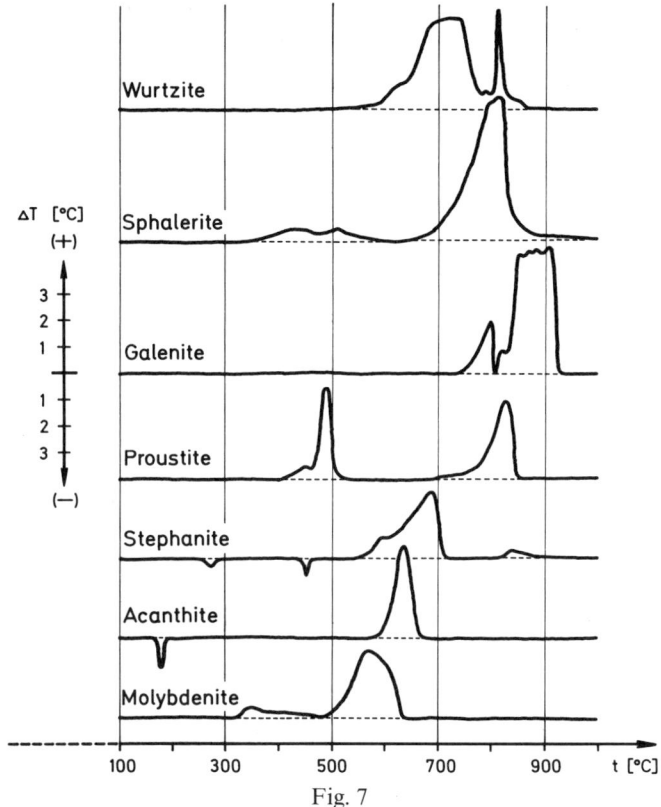

Fig. 7

or it is necessary to work in a sample holder wholly made of ceramic material. If only few sulfide samples are to be investigated, it may be sufficient to mix the substance carefully with inert material (twice the amount of sulfides) and use normal sample holders of metal. In any case, the proposed standard conditions of analysis (Table 1) must be varied if chalcogenides or sulfur are to be investigated by DTA.

The first DTA curves of sulfides and arsenides were published by HILLER and PROBSTHAIN (1955, 1956); SABATIER (1956); ASENSIO and SABATIER; LEVY; KULLERUD; and DUNNE and KERR (1961). MAUREL subdivided the chalcogenides according to the behaviour of the reaction products by the process of oxidation in five groups: 1. partly volatilizing of sulfur; 2. total volatilizing and oxidation of sulfur (or As, Se, Sb, Te); 3. oxidation into a sulfate (arsenate etc.); 4. total volatilizing of one component, remaining part of another as chalcogenide; 5. oxidation of all components (4 and 5 in the case of mixed S-As-etc original compounds). This subdivision and the DTA curves of Figs. 5, 6, and 7 demonstrate the numerous processes appearing during heating of chalcogenides. Frequently the formation of intermediate reaction products can be observed, the nature of which is not clear in every case. Numerous publications contain further DTA curves and data of chalcogenides. F. PAULIK, S. GAL, L. ERDEY, for instance, have determined small amounts of pyrite in Hungarian bauxites, CABRI has used DTA for the characterization of an unknown sulfide mineral, BARTON has studied the system Fe-As-S by means of DTA, HILLER and PROBSTHAIN have observed the transformation of chalcopyrites into a hitherto unknown γ-phase, which according to DTA runs by FRANZ can be lowered in its temperature from 530° (for pure $CuFeS_2$) to 480° C by partly substituting the sulfur by Se, and BOLLIN has studied the structural transformation of pyrrhotite.

A semi-quantitative determination of chalcogenides is much more difficult than of other minerals, because of their very strong and complicated oxidation behaviour. But the *chemical composition* of solid sulfide solutions can be determined by DTA where the minerals show a distinct structural transformation: the temperatures of inversion of Ag_2S and Cu_2S vary with different chemical composition (compare with III-2.3); substitution of S by Se lowers the transformation and oxidation peak temperatures in the series galena (PbS; $t_{ox}=800°$ C)—clausthalite, (PbSe; $t_{ox}=660°$ C, DUNNE and KERR, 1961), and substitution of Zn by Fe in sphalerites is also reflected by the lowering of a structural transformation temperature (KOPP and KERR, 1958a).

The data of Table 3 were obtained by heating (10°/min) a mixture of 50 mg sulfide (finely ground to 60–200 μ $\varnothing$) and 100 mg Al_2O_3 in the Mettler thermoanalyzer 2. The thermocouples of Pt-Pt_{90}/Rh_{10} had no contact with the samples. X-ray determinations showed the samples to be pure with the exception of germanite and bornite, which contained some chalcocite (peak at 101° C!). Furnace atmosphere: air (without turbulence); sample holder: platinum crucibles. Exactness of measurement: $\pm 1°$ C.

Endothermic deflections as occurring below 500° C in the curves of some Cu and Ag sulfides and of pyrrhotite reflect structural changes, exothermic deflections show the oxidation reactions going on in the form of several steps, and these oxidation reactions are generally very strong. The comparison of the author's data with those of other analysts show that the data of Table 3 agree with those of SABATIER; LEVY; ASENSIO and SABATIER; the data of MAUREL lie a little higher. The temperature of some structural transformations point out a certain relation to the type of structure, of such a kind that in the case of polymorphy the lower symmetric structure shows the lower temperature of structural changes. In the case of related chemical composition it is similar: the orthorhombic Cu-sulfides chalcocite and bournonite invert at lower

Chalcogenides 29

Table 3. DTA data of sulfides (in °C)

	Mineral, formula	Endothermic peaks (ΔT) structural transformation	Exothermic peaks (ΔT)					
(1)	Pyrite, FeS_2	580 (1.5) 639 (1.7)			468 (3.5)	515 (1.0)	570 (1.0)		
(2)	Marcasite, FeS_2		280 (0.2)	335 (0.2)	415 (6.0)	480 (1.0)	540 (3.5)	600 (3.3)	645 (0.5)	657 (3.0)
(3)	Pyrrhotite, FeS	370 (0.9)				490 (1.0)	570 (5.0)		
(4)	Arsenopyrite, FeAsS					490 (3.5)	540 (4.0)		
(5)	Chalcocite, Cu_2S	95 (0.4) 103 (1.0)				510 (3.5)			
(6)	Chalcopyrite, $CuFeS_2$		290 (0.4)		408 (5.7)	493 (1.0)	578 (4.5)		
(7)	Bornite, Cu_4FeS_5	101 0.3) Cu_2S-impurity!			429 (4.7)	496 (6.7)	581 (0.5)		
(8)	Bournonite, $2PbS \cdot Cu_2S \cdot Sb_2S_3$	137 (0.2)	288 (0.1)	327 (1.2)	432 (0.3)	490 (2.7)	578 (2.4)		
(9)	Tennantite, $Cu_3AsS_{3.25}$	319 (0.3) 753 (0.8) 815 (1.5) 850 (0.3)			437 (3.4)		555 (3.8)	600 (3.5)	
(10)	Germanite $Cu_3(Ge,Fe)S_4$	101 (0.4) Cu_2S-impurity!			448 (3.8)	477/493 (6.2)	570 (3.5)		
(11)	Acanthite, Ag_2S	180 (1.1)							644 (1.7)
(12)	Stephanite, $5 Ag_2S \cdot Sb_2S_3$	270 (0.3) 451 (0.6)						597 (0.8)	640 (3.5)	693 (2.5)	843 (0.3)
(13)	Proustite, Ag_3AsS_4				450 (0.5)	490 (3.5)					830 (3.0)

Table 3. (continued)

Mineral, formula	Endothermic peaks (ΔT) structural transformation	Exothermic peaks (ΔT)			
(14) Sphalerite, α-ZnS		420 (0.5)	505 (0.5)		820 (5.5)
(15) Wurtzite, β-ZnS				722 (3.5) 744 (3.5)	815 (3.8)
(16) Galena, PbS				688 (3.3) 854 (4.6) 874 (4.5) 885 (4.6)	910 (4.8)
(17) Molybdenite, MoS$_2$	345 (0.3)			570 (2.5)	

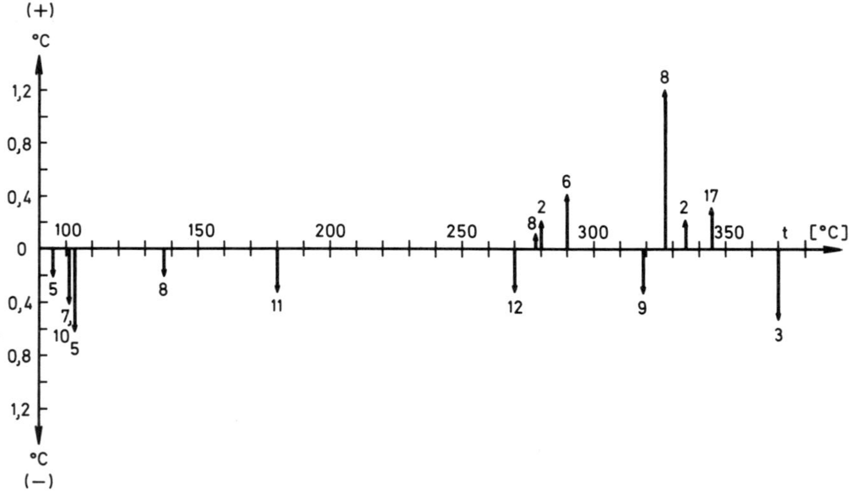

Fig. 8. Structural transformations (endothermic) and some oxidation reactions (exothermic) of sulfides <370° C; ordinate: ΔT(° C)

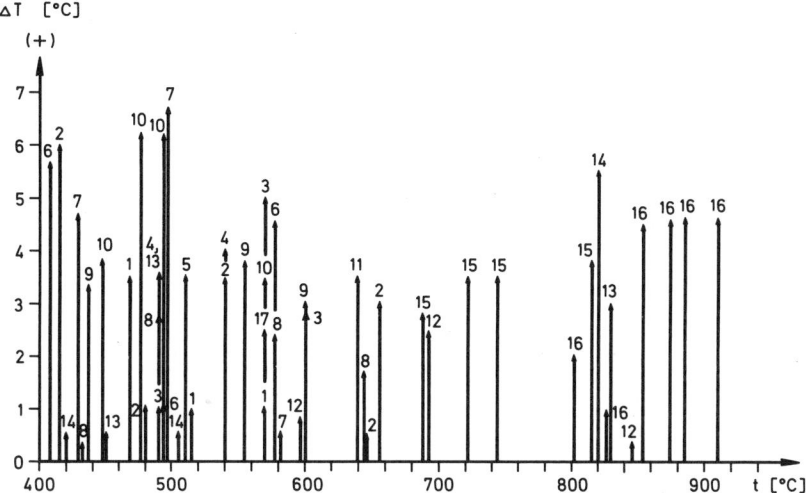

Fig. 9. Oxidation reactions of sulfides > 400° C; peak temperatures versus ΔT (ordinate). Numbers in Figs. 8 and 9 as in Table 3

temperatures (103° resp. 137° C) than the cubic tennantite (319° C), and the inversion temperature of the monoclinic acanthite (180° C) lies below that of the orthorhombic stephanite (270° C), both being silver sulfides.

2. Halogenides and Sulfates

2.1 Halogenides

The chlorides halite and sylvite can be identified differential thermal analytically at their melting temperatures. The DTA curves of the fluorides cryolite and fluorite show minor endothermic effects caused by structural changes (see Fig. 10). The cryolite is a suitable internal standard for the determination of the high-low inversion temperatures of quartz crystals (see III-8), because of its reversible structural transformation temperature of 562.7° C lying only 10° C below that of a *"normal"* quartz. Before the melting temperature of CaF_2 (1400° C), the DTA curve of fluorite shows two weak endothermic deflections reflecting structural changes, and one exothermic deflection somewhat stronger than the endothermic effects. The purpose of this exothermic effect is not clear. JÄGER and SCHILLING observed another exothermic deflection in DTA curves of fluorites from Wölsendorf/Bavaria, heated up to 600° C

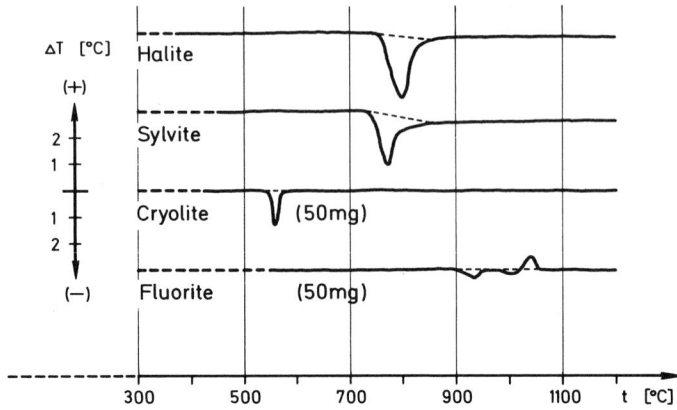

Fig. 10. DTA curves of halogenides; a halite, b sylvite, c cryolite, d fluorite from Wölsendorf

Table 4. DTA data of halogenides

Mineral, formula sample location	Amount of sample (mg)	Endothermic reactions (°C) ΔT(°C)	Exothermic reactions (°C) ΔT (°C)
Halite, NaCl Staßfurt, Germany	100 + 50 Al_2O_3	801 ± 1; 2.5	
Sylvite, KCl Staßfurt, Germany	100 + 50 Al_2O_3	775 ± 1; 2.0	
Cryolite, Na_3AlF_6 Ivigtut, Greenland	50	562.7 ± 0.3; 1.3	
Fluorite, CaF_2 Wölsendorf, Bavaria	50	932; 0.2 1010; 0.1	1040; 0.5
Fluorite Andreastal, Harz, Germany	50	928; 0.2 1007; 0.1	1032; 0.6

between 400 and 600° C, which has not been found in the author's DTA runs of fluorites from Wölsendorf as well as from other localities. The exothermic effect in the curves of JÄGER and SCHILLING may have been caused by small amounts of impurities (organic matter or sulfides).

The data of Table 4 were obtained by standard conditions, with the exception of the amounts of sample of fluorides (50 mg instead of 100 mg) and mixing of the chlorides with 50 mg Al_2O_3, so preventing the molten chlorides from glueing on the crucible.

DTA data of rare halogenide minerals are contained in the publications of KOHLS and RODDA (iowaite, 4 Mg(OH)$_2$ FeOCl × H$_2$O) and of

VAN VALKENBURG and RYNDERS (cuspidine, CaO 2SiO$_2$ CaF$_2$, melting point at 1405° C). HEIDE and BRÜCKNER demonstrated in some salt systems that DTA can be taken for the distinction of systems with solid solutions and systems with eutectic melting character, provided that they are analysed at a low heating rate, because the equilibrium will be attained very slowly.

2.2 Sulfates

After BERG (1970) heat changes during heating of sulfates can be caused by numerous processes like dissociation, dehydration, structural transformation, melting (all being endothermic), transition from metastable to more stable intermediate reaction products, crystallization, oxidation (exothermic). In many cases DTA curves will result with ten or more partly overlapping deflections which can be directed reciprocally.

The interpretation of these numerous deflections will not be very simple in any case. HEIDE, who has studied the thermal behaviour of several salt minerals and rocks (HEIDE, 1962–1967, 1968, 1969), contrary to BERG (1961, 1970), comes to the conclusion that a phase analysis of hydrated sulfates by means of DTA is impossible; during heating of these sulfates on 100° C melts will form which will lose more and more water by evaporation with increasing temperature. This results in shifting of the equilibrium and new sulfates, containing less water, will precipitate as intermediate products from oversaturated "solutions". That means: by heating of hydrated sulfates, transformations and reorganizations of the components appear. The thermal effects being found at higher temperatures cannot be traced back to the original substances, but only to the intermediate reaction products of them (HEIDE, 1962). These intermediate products are often X-ray amorphous compounds which crystallize at higher temperatures (= exothermic effects in DTA curves of hydrated sulfates).

Such complicated processes of dehydration connected with the formation of intermediate reaction products impoverished in water can for instance be observed in the behaviour of hydrated magnesium sulfates (HEIDE, 1962–1967, BERG, 1970). The original Mg-sulfate mineral is epsomite, containing seven H$_2$O.

$$MgSO_4 \cdot 7H_2O \xrightarrow{52°} MgSO_4 \cdot 6H_2O \xrightarrow{94°} MgSO_4 \cdot 4H_2O \xrightarrow{106°, 117°} MgSO_4 \cdot 3H_2O \xrightarrow{138°} MgSO_4 \cdot 2H_2O \xrightarrow[225°]{164°, 184°} \underset{\text{(kieserite)}}{MgSO_4 \cdot H_2O} \xrightarrow{340°} MgSO_4$$

The DTA curve of an original epsomite therefore shows nine endothermic events. Other examples for such "polythermic dehydrations"

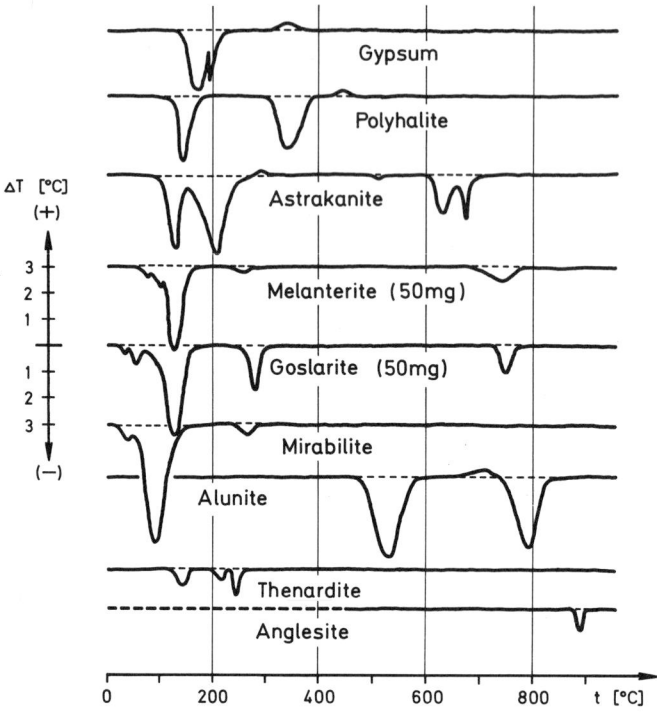

Fig. 11. DTA curves of some sulfates

(HEIDE, 1969) are the dehydration of pickeringite, $MgAl_2(SO_4)_4 \cdot 22\,H_2O$ (BROUSSE and GUERIN; JENNI) or of goslarite, $ZnSO_4 \cdot 7\,H_2O$ (see Fig. 11). But there are also hydrated sulfates showing only few thermal effects, for instance the sulfates rich in water melanterite, $FeSO_4 \cdot 7\,H_2O$, which transforms between 85 and 100° C by losing $3\,H_2O$ into the rare mineral rozenite (see BROUSSE, GASSE-FOURNIER and LEBOUTEILLER), loeweite, $4\,Na_2SO_4 \cdot 4\,MgSO_4 \cdot 9\,H_2O$ (HEIDE, 1966; VON HODENBERG et al.), and the poorly hydrated minerals gypsum, $CaSO_4 \cdot 2\,H_2O$ (LJUNGGREN; PIÈCE; LEHMANN and HOLLAND), polyhalite, $K_2Ca_2Mg(SO_4)_4 \cdot 2\,H_2O$ (HEIDE), bloedite, $Na_2Mg(SO_4)_2 \cdot 4\,H_2O$ and other double salts (HEIDE, 1968; FÖLDVARI-VOGL).

Among the hydrated sulfates the most frequent mineral gypsum has to be investigated most often. POWELL proved fewer than 2 weight-% of gypsum in rock samples by means of DTA which could not be found by X-ray, SMYKATZ-KLOSS (1966) attained clear DTA deflections in the curves of mixtures of 99.8 weight-% of Al_2O_3 and 0.2% of gypsum, with an apparatus of higher sensibility, GUTT and SMITH found the hitherto

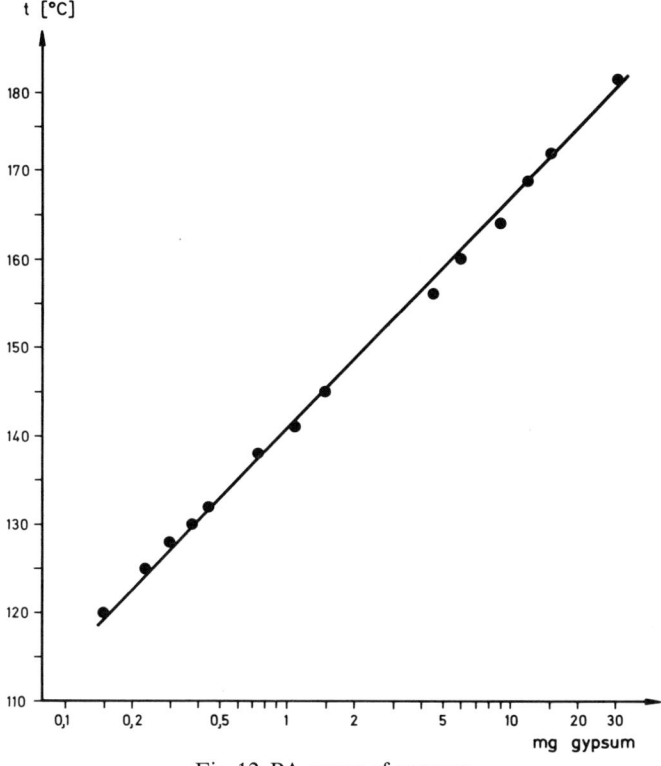

Fig. 12. PA-curve of gypsum

unknown α-phase of $CaSO_4$ at more than 1000° C. According to the PA-curve (Fig. 12), small amounts of gypsum can be determined semi-quantitatively in soil or sediment samples (compare with SMYKATZ-KLOSS, 1966). Table 5 contains the DTA data of some sulfates obtained under standard conditions (see Table 1), with the exception of the sample holder: the nickel block here used has a higher selectivity for thermal effects lying close together. The amounts of gypsum, melanterite and goslarite differ from the standard amount of 100 mg, the first in order to show the dependence of the peak temperatures on the amount of sample, the second and third because of shortness of substances.

The preparation of samples has to be very careful, since some hydrated sulfates already lose their water of crystallization by mechanical grinding in a mortar.

Among the sulfates which contain no water or which only incorporate OH^- into their structure, some can be identified well differential thermal analytically. They show dehydration peaks similar to those of

Table 5. DTA data of some sulfates (in °C)

Mineral, formula	Amount of sample (mg)	Endothermic reactions (ΔT)	Exothermic reactions (ΔT)
Gypsum, $CaSO_4 \cdot 2H_2O$	0.15	120 (0.02)	
Gypsum, $CaSO_4 \cdot 2H_2O$	0.3	128 (0.05)	
Gypsum, $CaSO_4 \cdot 2H_2O$	10	151 (0.3), 164 (0.5)	
Gypsum, $CaSO_4 \cdot 2H_2O$	75	175 (1.8), 198 (1.6)	340 (0.3)
Gypsum, $CaSO_4 \cdot 2H_2O$	150	178 (2.6), 200 (2.2)	342 (0.5)
Polyhalite, $K_2Ca_2Mg(SO_4) \cdot 2H_2O$	100	145 (2.5), 342 (2.0)	443 (0.2)
Astrakanite, $Na_2Mg(SO_4)_2 \cdot 4H_2O$	100	127 (2.8), 208 (3.0), 507 (0.1), 635 (1.5), 673 (1.8)	295 (0.2)
Melanterite, $FeSO_4 \cdot 7H_2O$	50	83 (0.3), 110 (0.5), 130 (3.2), 310 (0.2), 735 (0.5)	
Goslarite, $ZnSO_4 \cdot 7H_2O$	50	35 (0.3), 55 (0.5), 137 (3.5), 288 (1.7), 752 (1.0)	
Mirabilite, $Na_2SO_4 \cdot 10H_2O$	100	40 (0.6), 97 (4.5), 266 (0.4)	
Alunite, $KAl_3(SO_4)_2(OH)_6$	100	537 (3.0), 795 (2.7)	715 (0.3)
Thenardite, Na_2SO_4	100	147 (0.6), 215 (0.4), 245 (1.0), 890 (0.8)	

clay minerals between 500 and 600° C (e.g. the alunite, see KASHKAI and BABAEV and Table 5), or structural transformations like anglesite (895° C, McLAUGHLIN) or thenardite which can be recognized at three reversible structural changes, at 147, 215, and 245° C (see Fig. 11).

By means of combined thermal optical methods, as has been described by HEIDE (1969) for the examples epsomite and kieserite, in favourable cases the intermediate reactions appearing on heating of hydrated sulfates allow the determination of kinetic data about these reactions.

3. Oxides and Hydroxides

3.1 Oxides

Oxidic minerals can only be recognized differential thermal analytically by the appearance of structural or magnetic changes. The high-low inversion of quartz ($\sim 573°$ C) has been for many years a means of calibrating the temperature, as well as for calorimetric measurements (following a proposal of FAUST), but neither the inversion temperature ($= t_i$) nor the heat of reaction (3.1 cal/g after FAUST) can be used for calibration, since the t_i can vary by a great temperature interval (KEITH and TUTTLE; SMYKATZ-KLOSS, 1970, see III-8), and the heat of reaction is dependent on the degree of disorder of the crystals. GRIM and ROWLAND have measured 565° for the t_i of authigenically formed quartz crystals of soils. The t_i-variation of the SiO_2 minerals cristobalite and tridymite is still much greater than the t_i-variation of quartz crystals (see III-7). Magnetites oxidize in two steps to Fe_2O_3, reflecting two very strong exothermic effects with peaks at 380 and 580° C. Shape and temperatures of these peaks are dependent on the grain size (EGGER), and on some other factors (KIRSCH). A weak endothermic effect at 680° C points out the Curie point of Fe_2O_3 (EGGER); the temperature of this peak varies with different contents of titanium and manganese (compare with III-3). In DTA curves of hematite PETERS obtained a broad endothermic deflection between 400 and nearly 700° C with a small endothermic peak at 675° C, which probably points out the Curie point of the Fe_2O_3. In a lot of the author's DTA runs of hematite only this small endothermic effect (at 675–680° C) has been observed, but only with large amounts of sample material (>80 mg hematite). The Mn-oxides hausmannite and pyrolusite can be recognized by several structural changes (KULP and PERFETTI; AGIORGITIS). Results of Table 6 are obtained according to standard conditions. DTA curves see III-7, -8, -9.

Table 6. DTA data of some oxides (in °C)

Mineral	Formula	Endothermic reactions (ΔT), cause		Exothermic react. (ΔT)
Quartz	SiO_2	520–578 (0.3–2)	Inversion	
Cristobalite	SiO_2	90–270 (0.3–2)	Inversion	
Tridymite	SiO_2	100; 150 (0.2–1)	Inversion(s)	
Hematite	α-Fe_2O_3	675 (0.1–0.3)	Curiepoint	
Magnetite	$FeO.Fe_2O_3$	450–590 (0.1–0.3)	Curiepoint	300–450; 480–700 (very strong)
Pyrolusite	β-MnO_2	690	Inversion	
Hausmannite	$MnMn_2O_4$	750 (0.2)	Inversion	
Maghemite	γ-Fe_2O_3	640 ± 5 (0.2)	Curiepoint	

3.2 Hydroxides

DTA curves of hydroxides have been published by ORCEL, NORTON (1939b), GRIM and ROWLAND, KULP and PERFETTI, KULP and TRITES, MACKENZIE (1952), KELLY, PETERS, SMYKATZ-KLOSS (1966), BUURMAN and VAN DER PLAS (1968), KELLER, M.G. WILSON et al. and others. Contrary to the Fe-hydroxides, the Al-hydroxides can be recognized and

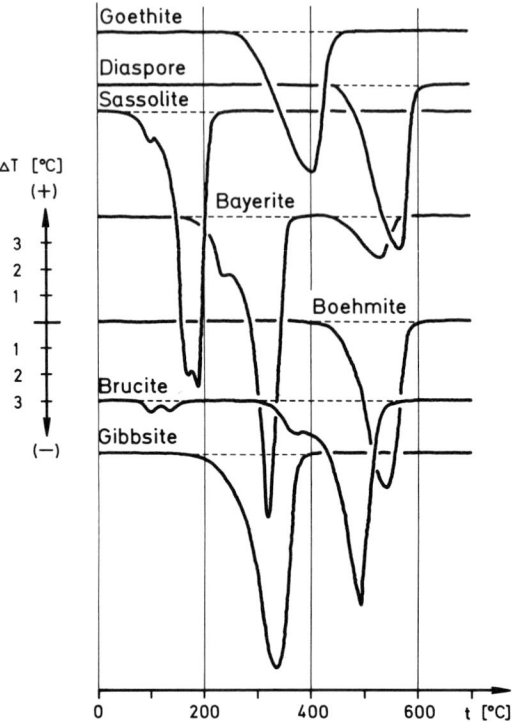

Fig. 13. DTA curves of some hydroxides

differentiated very well (Table 7, Fig. 13). The results of MACKENZIE (1952) and of KULP and TRITES demonstrate that the mineral goethite, α-FeOOH, is more stable than lepidocrocite. (The dehydration and decomposition peak temperature of the first one is higher.) But KELLY has shown that this statement is only valid for well-ordered goethites: and there are many fine-grained, often X-ray amorphous goethites in soils which are clearly disordered and dehydrating at 50° or more *below* the dehydration temperature of lepidocrocite. Some of these disordered minerals which have been identified exactly as goethites by X-ray methods also show the exothermic effects, which after MACKENZIE (1952) should be characteristic for β-FeOOH and lepidocrocite. These exothermic deflections, which appear for instance in the DTA curve of the "limonite" sample, are a further criterion of disorder of hydroxide structures; they reflect the (re-)cristallization of badly crystallized, strongly disordered minerals. The "limonite" of Table 7 is goethite, very fine-grained and strongly disordered. Besides the degree of crystallization,

Table 7. DTA data of hydroxides (in °C)

Mineral	Formula	Endothermic reactions (ΔT)	Exothermic reactions (ΔT)
Goethite	α-FeOOH	411 (5.4)	
Al-goethite (30 mole-% Al; synthetic, THIEL)	α-(Fe,Al)OOH	372 (3.4)	
Lepidocrocite	γ-FeOOH	345 (4.6)	470 (5.2)
Gibbsite (hydrargillite)	γ-Al(OH)$_3$	340 (8.2)	
Böhmite	γ-AlOOH	545 (6.3)	
Böhmite (from ageing of gibbsite)	γ-AlOOH	526 (2.2)	
Diaspore	α-AlOOH	572 (6.2)	
Bayerite	α-Al(OH)$_3$	233 (1.0)[a]; 325 (11.0); 535 (1.5)[a]	
Manganite	γ-MnOOH	370 (4.4)	
Brucite	Mg(OH)$_2$	98 (0.35); 135 (0.3); 374 (0.2); 493 (7.5)	
Sassoline	B(OH)$_3$	102 (0.5); 172 190 } (10.0)	
"Limonite"	(Fe$_2$O$_3 \cdot$ H$_2$O)	340 (3.7)	423 (5.0) 475 (5.5)

[a] Impurity of less ordered goethite (233°!) and diaspore (535°).

the chemical composition also influences the dehydration of goethites: a synthetic goethite sample which had incorporated nearly 30 mole-% AlOOH in its structure (THIEL) showed a peak 40° C lower and a broader deflection of lower ΔT compared with the peak of a well-ordered and pure goethite.

Recently KÜHNEL, VAN HILTEN and ROORDA studied the crystallinity (the degree of disorder) of goethites from laterite profiles by means of X-ray and DTA methods. Their DTA curves and data agree well with the results of the author (compare with III-1.5).

The dehydration temperature of boehmites and other hydroxides, too, are influenced by the degree of crystallization or disorder. The one boehmite formed by ageing, fine-grained and characterized by broad X-ray reflections (Guinier diagram) dehydrated 20° earlier than a well-crystallized sample. Both the Al(OH)$_3$-minerals gibbsite and bayerite scarcely differ in DTA curves. Data of Table 7 are obtained according to standard conditions.

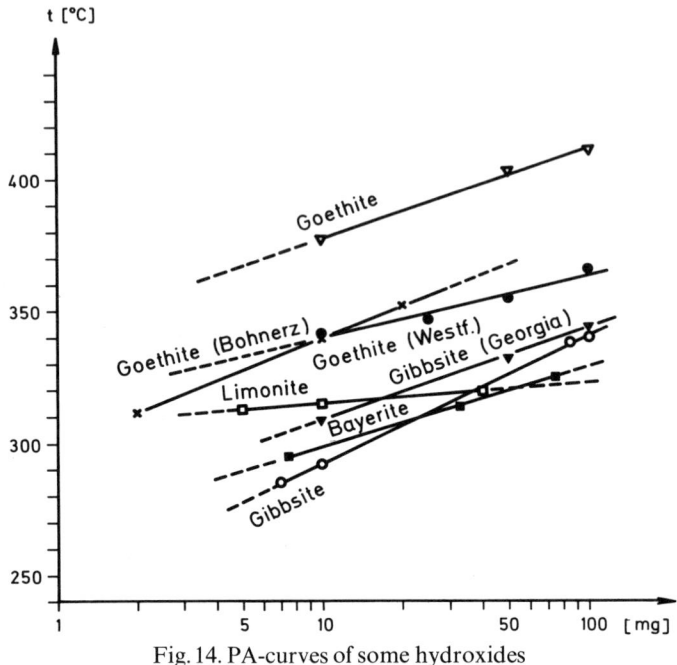

Fig. 14. PA-curves of some hydroxides

3.3 Soils and Iron Ores

Soils are weathering products not having been transported. Their main components are clay minerals (kaolinite, montmorillonite and so on), and hydroxides of iron and aluminum. These hydroxides are mostly badly crystallized and generally very fine grained. Some of them are used as iron ores (brown ore, bean ore). The "braune Glaskopf" (Table 8) consists of hematite and nearly 5% badly crystallized lepidocrocite, the "laterite" from Vogelsberg of gibbsite and well-ordered kaolinite (on order-disorder of kaolinites see III-6), the bauxite from Georgia consists of a mixture of gibbsite (>95%) and diaspore; the bean ore from Tuttlingen consists of well-crystallized goethite (~20%), detrital quartz (~20%) and well-ordered kaolinite, the oolitic iron ore from Westphalia mainly of badly-crystallized goethite and a little well-ordered kaolinite, and lastly the "white iron ore" from Sulzbach-Rosenberg/Bavaria of detrital quartz, some disordered kaolinite and two goethites different in their degree of crystallization (peaks at 262° and 370° C). The coexistence of two different goethites is not unusual (KELLY). On the term "degree of crystallization" see III-7). Data of Table 8 according to standard methods (Table 1).

Table 8. DTA data of soils and iron ores (in °C)

Sample	Locality	Endothermic reaction (ΔT)	Exothermic reaction (ΔT)
Brown glasshead	Harz, Germany	227 (0.3)	
Laterite	Hungen/Vogelsberg Germany	307 (2.0) 568 (1.2)	
Bauxite	Georgia	344 (10.1) 559 (0.3)	
Bean ore	Tuttlingen, Germany	352 (0.6) 558 (3.8), 573 (0.8)	
Iron oolite	Neeßen, Westfalia, Germany	366 (2.8) 565 (0.7)	990 (0.7)
White iron ore	Sulzbach-Rosenberg, Bavaria	262 (1.6) 370 (1.7) 526 (0.4) 576 (0.5)	985 (0.3)

4. Carbonates and Nitrates

Up to 1970 nearly 350 DTA publications on carbonates have been published (after WEBB and KRÜGER). A survey of the differential thermal analytical behaviour of the most important carbonate minerals is given by CUTHBERT and ROWLAND, BECK (1950), FÖLDVARI-VOGL, SPOTTS and SMYKATZ-KLOSS (1964). Besides the description of the DTA characteristics of several minerals (e.g. cerussite: WARNE and BAYLISS; huntite: FAUST, 1953; hydromagnesite: CAILLÈRE, 1943a,b; carletonite: CHAO; manasseite: ROSS and KODAMA etc.), above all the Ca-Mg-Fe-Mn carbonates have been studied over and over again, since the analysis of these minerals involves a lot of problems reflected in major differences of their DTA curves.

The greatest differences (discrepancies) have been described on siderites and ankerites (FREDERICKSON; ROWLAND and JONAS; KULP, KENT and KERR; STALDER; TRDLIČKA, 1966; DASGUPTA a.o.), the DTA curves of which showed or did not show the exothermic peak at temperatures of 400 or 700° C caused by the oxidation of $Fe^{II} \rightarrow Fe^{III}$. The author's DTA runs under varying heating conditions demonstrated that the main purpose of these discrepancies in the DTA curves of Fe-carbonates is the *heating-rate* used: at heating rates below 10°/min only the exothermic deflection appears in the DTA curve of siderites, first at heating rates of more than 10°/min both effects (the decomposition and the Fe^{2+}-oxidation) will be selected in the curve. Very different DTA data exist from the

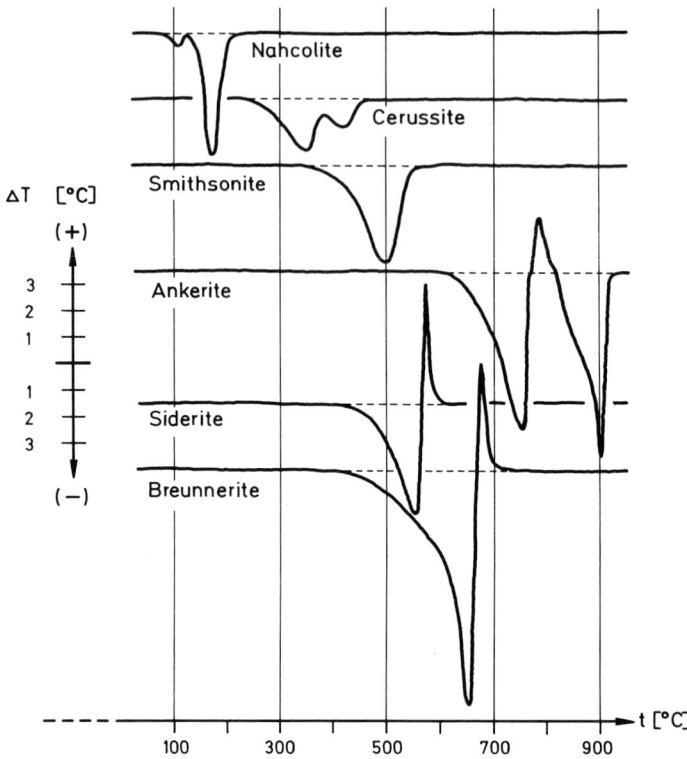

Fig. 15. DTA curves of nahcolite, cerussite, smithsonite, ankerite, siderite, and breunnerite

Mn-carbonates rhodochrosite, Mn-calcite and kutnahorite, too (FRONDEL and BAUER; TRDLIČKA, 1964; TSUSUE a. o.).

The DTA characteristics of dolomites are also strongly variable (shape, temperatures and ΔT of deflections): low Fe contents in the dolomite structure (~ 0.5 weight-%) will cause a third endothermic peak (see Fig. 16), traces of impurities of NaCl lower the peak temperatures by nearly 30° C (GRAF), and other crystal chemical differences (e.g. the incorporation of Mn into the structure) influence their thermal behaviour (ROWLAND and BECK; HAUL and HEYSTEK; BRADLEY, BURST and GRAF). All these statements lead to the consequence that reproducible data in DTA investigations can only be obtained by working under highly standardized conditions of analysis. For this study 45 samples (35 carbonate minerals) were available; each sample was checked by X-ray analysis on existing impurities which had been taken into consideration

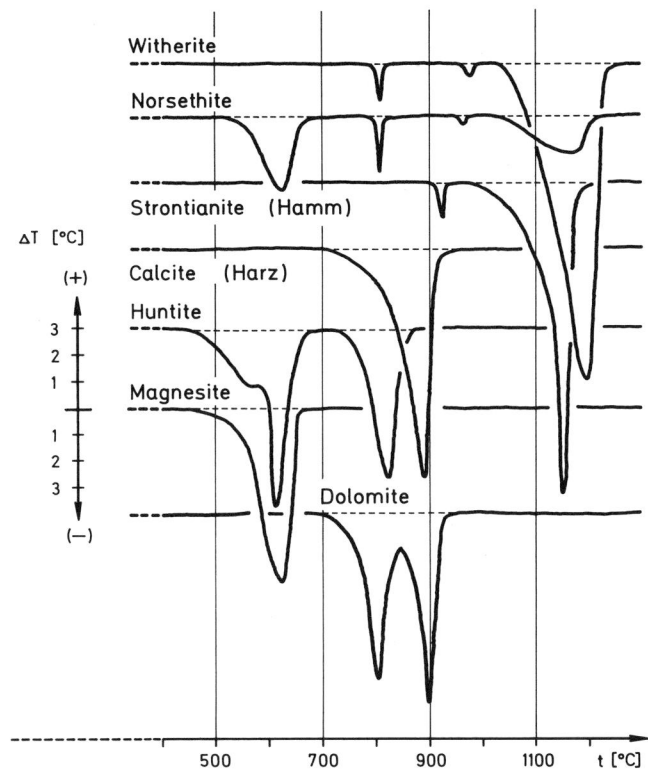

Fig. 16. DTA curves of witherite, norsethite, strontianite, calcite, huntite, magnesite, and dolomite

in the drawings of PA-curves. Classification and formulas of carbonate minerals after STRUNZ.

4.1 Carbonates Free of Water and without Other Anions

The minerals of this group can be recognized differential thermal analytically by their decomposition temperatures (Table 9). Some of them decompose at very high temperatures ($>1000°$ C) which cannot be reached in a conventional DTA apparatus; but by construction of the PA-curves these minerals (strontianite, witherite, norsethite) can also be determined by means of DTA running up to $1000°$ C (Fig. 19, compare with SMYKATZ-KLOSS, 1967a and b). A semi-quantitative determination of all carbonate minerals mentioned in this study will be possible by

Table 9. DTA data of carbonates free of water without strange anions (in °C)

Mineral, formula	Sample locality	Endothermic react. 100 mg (ΔT)	1 mg (stand. t.)	Exothermic reaction (ΔT)
Nahcolite, $NaHCO_3$	Searles Lake, Calif.	110 (0.5) 170 (4.6)	128	
Cerussite, $PbCO_3$	Ems, Nassau, Germany	350 (1.9) 427 (1.0)	298 343	
Cerussite	Unknown	340 (3.0) 407 (1.0)	285 335	
Mn-calcite, $(Ca,Mn)CO_3$	Freiberg, Saxony	492 (0.3) 925 (0.7)		530 (0.3)
Smithsonite, $ZnCO_3$	Tsumeb, Namibia	499 (3.7)	432	
Siderite, $FeCO_3$	Siegen, Germany	555 (4.2)		572 (4.6)
Siderite	Wissen/Sieg, Germany	540 (8.5)		571 (7.5)
Huntite, $CaMg_3(CO_3)_4$	Tea tree galley, S-Australia	568 (2.0) 613 (6.8) 826 (5.7)	489 540 647	
Magnesite, $MgCO_3$ (dense)	Euboea, Greece	625 (6.6)	527	
Magnesite (spaty)	Hoboken, N.J.	643 (7.0)	541	
Breunnerite, $Mg_{0.45}Fe_{0.55}CO_3$	Tonberg, Salzburg, Austria	654 (9.0)		677 (4.0)
Ankerite, $Ca(Fe,Mg)(CO_3)_2$	Kohlenbach bei Siegen, Germany	762 (6.0) 898 (7.1)		782 (2.0)
Ankerite (8.5 weight-% FeO; 30.95% CaO)	Kohlenbach bei Siegen, Germany	773 (5.5) 882 (5.0)		785 (1.0)
Dolomite, $CaMg(CO_3)_2$	Natural Bureau of Standards, sample 88	807 (6.3) 901 (7.0)	753	
Dolomite	Rothenzechau, Germany	807 (7.5) 897 (10.5)	750	
Plumbocalcite, $(Ca,Pb)CO_3$ [0.4 weight-% PbO]	Bleiberg, Austria	890 (10.7)	710	
Calcite, $CaCO_3$	Unknown	895 (11.4)	712	
Calcite	St. Andreasberg/Harz, Germany	898 (8.9)	702	
Strontianite, $SrCO_3$	Hamm, Westfalia, Germany	929 (1.3) 1142 (10.8)		
Strontianite	Ahlen, Westfalia	925 (1.4) 1148 (12.3)	905	
Strontianite	Drensteinfurt, Westfalia	925 (1.4) 1151 (10.0)	910	
Strontianite	Loch Strontian, Scotland	928 (1.4) 1151 (10.6)	910	

Table 9. (continued)

Mineral, formula	Sample locality	Endothermic react. 100 mg (ΔT) 1 mg (stand.t.)	Exothermic reaction (ΔT)
Norsethite, BaMg(CO$_3$)$_2$	Synthetic (LIPPMANN)	628 (2.8) 810 (2.1) 969 (0.3) 1174 (1.4)	925
Witherite, BaCO$_3$	Alston Moore, England	810 (1.4) 981 0.5) 1195 (12.0)	940

means of the indirect DTA characteristics PA-curve and standard peak temperature, with the exception of the Fe-Mn-carbonates siderite, ankerite, breunnerite, rhodochrosite (compare with I-2.5.2, see Figs. 17–19, 21 etc.). In DTA curves of Fe-Mn-carbonates, both the endothermic

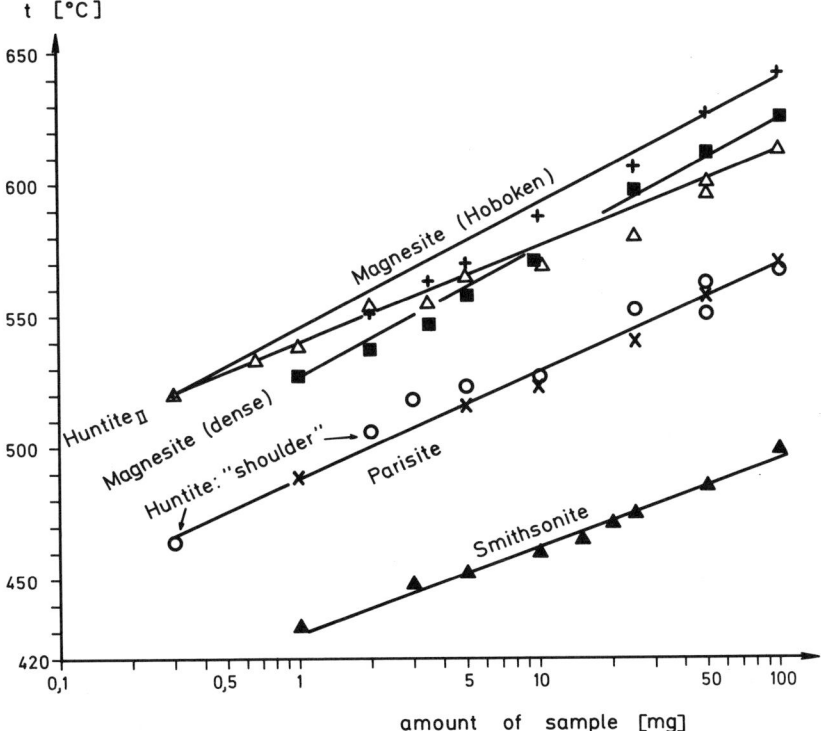

Fig. 17. PA-curves of magnesite, smithsonite, huntite, and parisite

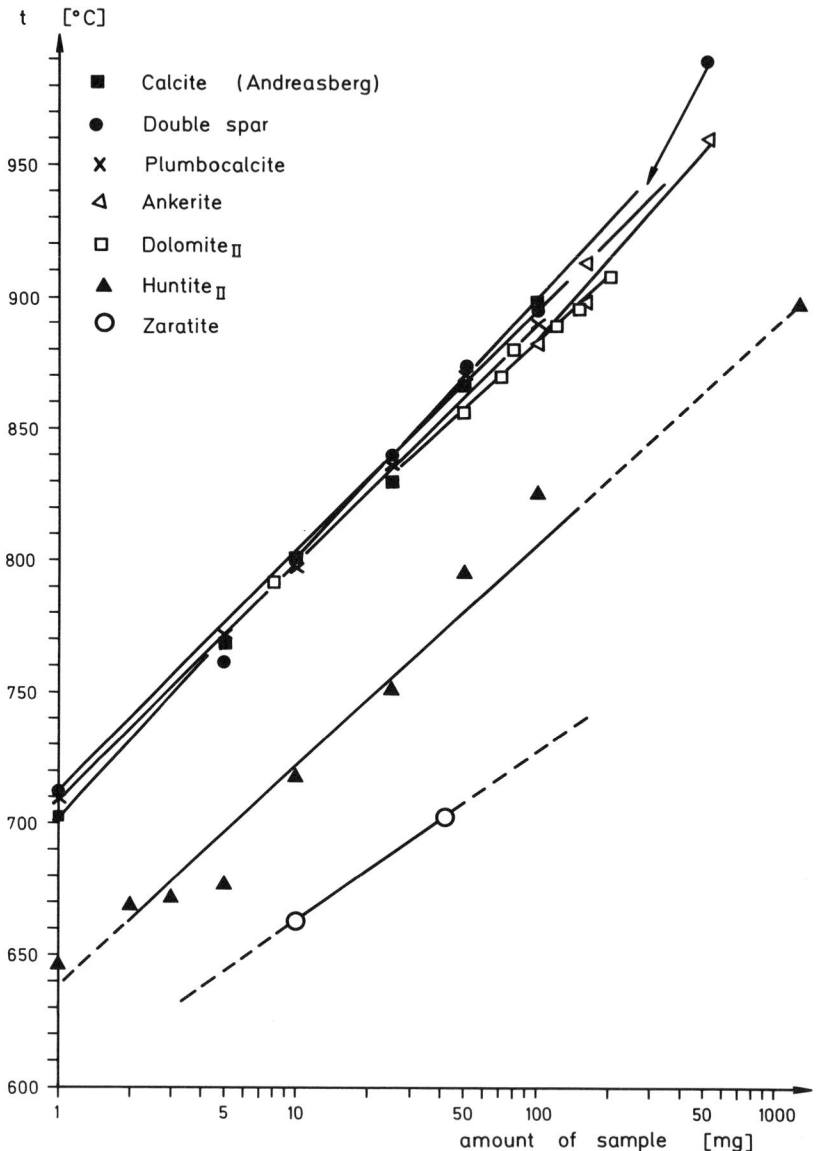

Fig. 18. PA-curves of calcite, dolomite, ankerite, and zaratite

decomposition peak and the immediately following exothermic peak caused by the oxidation of Fe^{2+} and Mn^{2+} overlap and intrude, especially in small amounts of material. This prevents the construction of

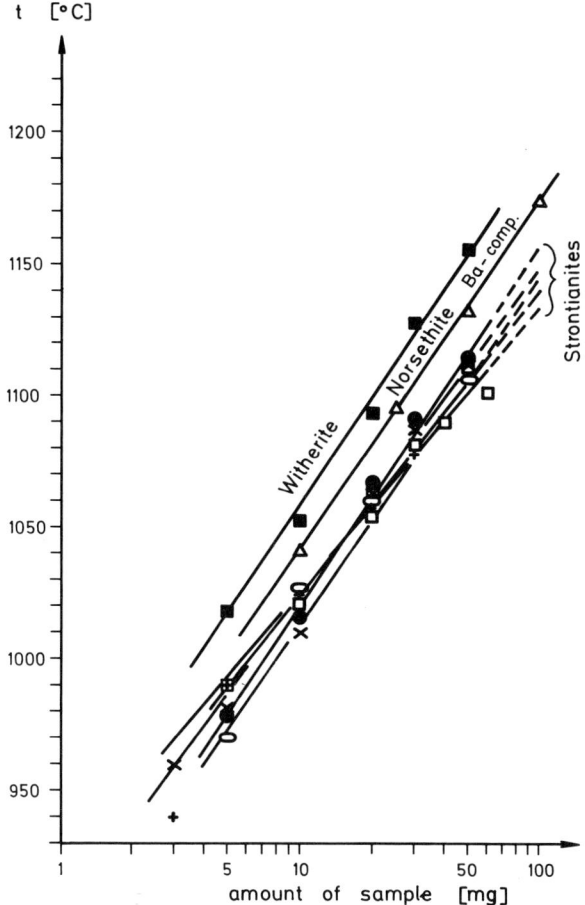

Fig. 19. PA-curves of strontianite, witherite, and norsethite

PA-curves being exact enough to determine the minerals semi-quantitatively. The data of Table 9 were obtained by standard methods, with the exception of the sample holder: the nickel block used allows a higher (better) selectivity of effects lying close together than a Pt-sample holder. Exactness of measurements: $\pm 1°$ C.

Some of the carbonate minerals of this group also show structural transformations at characteristic temperatures. The Ba-carbonates witherite and norsethite have transformations from orthorhombic to hexagonal at $811 \pm 2°$ C, and from hexagonal to cubic at $969 \pm 1°$ (norsethite, SMYKATZ-KLOSS, 1967b), resp. $981 \pm 1°$ (witherite). These temperatures of structural transformations are suitable for calibrating a DTA apparatus.

Table 10. Temperatures of the monotropic transformation aragonite → calcite

Mineral, variety	Sample locality	Transformation temperature, °C ± 1°
Clear aragonite-thrilling	Calanes, Aragone, Spain	447–456 (8 runs)
Pisolite	Karlsbad, CSSR	390
Fibrous aragonite	Karlsbad, CSSR	387
Tarnowitzite (Pb-aragonite)	Tarnowitz, Silesia, Poland	447
10 Aragonite samples according to Faust (1950)		387, 400, 447 (2 ×), 458, 461, 462, 477, 484, 488
Aragonite, according to Peters (1962)	Switzerland	410

Orthorhombic strontianite transforms into hexagonal $SrCO_3$ between 880 and 930° C.

Petrogenetic implications of frequent carbonate rocks can possibly be obtained by studying the conversion behaviour of the $CaCO_3$-modifications aragonite and calcite (investigations on this problem have been started). The following temperatures of conversion of aragonite into calcite have been determined for 1 atm. pressure (Table 10).

Small amounts of Ba and Sr (more than 0.3 weight-%) incorporated into the aragonite structure can be recognized by minor endothermic effects immediately following the $CaCO_3$-decomposition peak. The temperatures of these little peaks can be taken for the determination of the strontium contents of these aragonites (III-1.2). The PA-curves of Fig. 18, 21, 24, 26, and 27 contain some values from a publication of Beck (1950, all values representing sample amounts of more than 100 mg). Though Beck used a different DTA apparatus and large amounts of sample material (generally 500 mg), his data can be compared very well with the author's results. This is demonstrated by the fact that the PA-curves are also valid for Beck's data: this concordance was only possible because the conditions of analysis of both authors were similar.

4.2 Carbonates Free of Water with Other Anions

The seven studied minerals of this group contain the anions OH^-, F^- or Cl^- besides CO_3^{2-}. Methods of analysis as in 4.1, DTA curves and PA-curves see Figs. 20 and 21. Exactness of measurements: ± 1° C.

The shape of the endothermic deflections caused by release of OH^-, F^-, or Cl^- as well as by decomposition of the structures is well-rounded

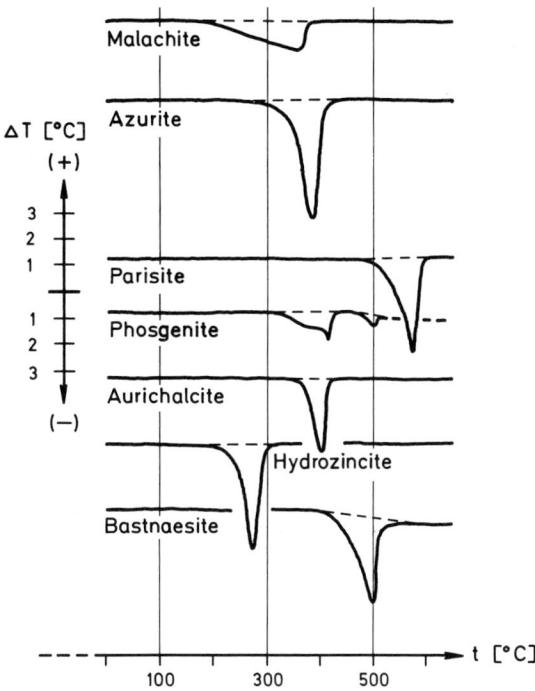

Fig. 20. DTA curves of carbonates free of water but with other anions

Table 11. DTA data of carbonates free of water with strange anions (in °C)

Mineral, formula	Sample locality	Decomposition temperature (ΔT) 100 mg	1 mg (stand. temp.)
Malachite $Cu_2[(OH)_2/CO_3]$	Katharinenburg, Ural, UdSSR	366 (7)	305
Azurite $Cu_3[(OH)_2/CO_3)_2]$	Bisbee, Arizona	390 (4.5)	336
Hydrozincite-I $Zn_5[(OH)_3CO_3]_2$	Santander, Spain	267 (7)	254
Hydrozincite-II	Santander, Spain	278 (4)	262
Aurichalcite, $(Zn,Cu)_5[(OH)_3/CO_3]_2$	Tsumeb, Namibia	405 (2.8)	365
Bastnaesite, $Ce[Fe/CO_3]$	Carona, Mexico	500 (3.5)	458
Parisite, $CaCe_2[F_2(CO_3)_3]$	Muzo, Columbia	570 (3.5)	489
Phosgenite, $Pb[Cl_2/CO_3]$	Sardinia	415 (1), 500[a]	412 (!)

[a] Melting of the remaining Pb-oxy-chloride after decomposition of the carbonate.

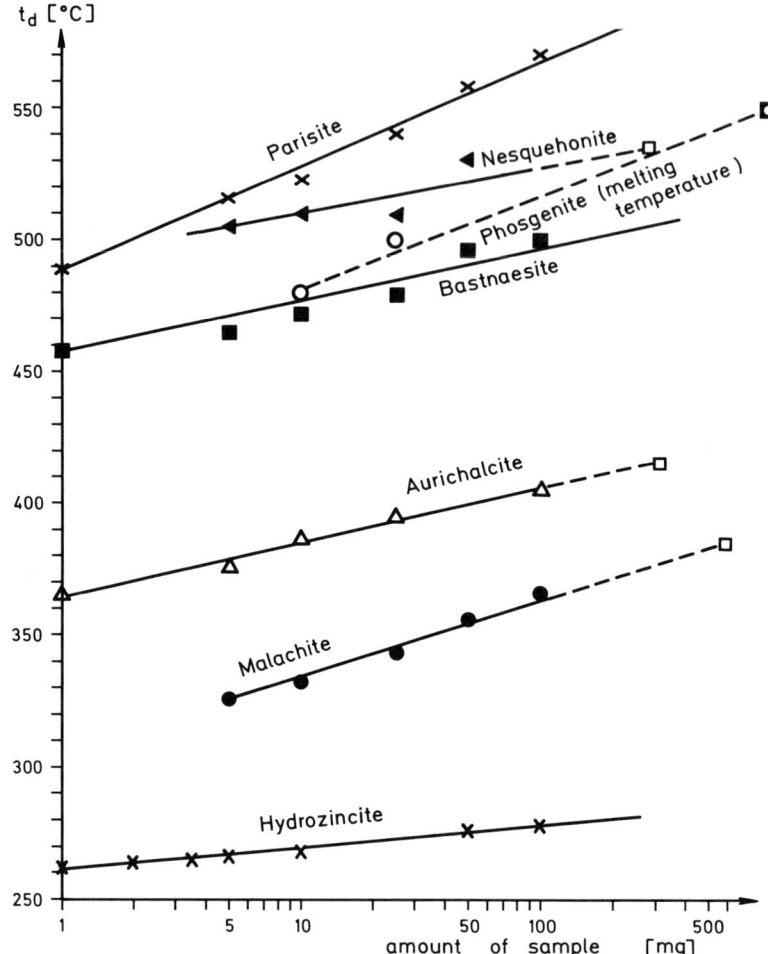

Fig. 21. PA-curves of carbonates free of water but with other anions

(contrary to the shape of the DTA peaks of the minerals of group 4.1, compare with Figs. 15, 16, and 20). The DTA curves of aurichalcite and bastnaesite show both endothermic effects selected (separated), but only in samples containing fewer than 25 mg bastnaesite or aurichalcite.

4.3 Hydrated Carbonates without Other Anions

Table 12 and Figs. 22–24 contain the DTA data and DTA curves of the six minerals of this group. Dehydration of the minerals nesquehonite, soda, trona, pirssonite and gaylussite appears in several steps. DTA

Table 12. DTA data of hydrated carbonates without strange anions (in °C)

Mineral, formula	Sample locality	Endothermic reactions (ΔT)			Exothermic reactions (ΔT)
		dehydration	decomposition	Stand. temper.	
Nesquehonite (50 mg!), $MgCO_3 \cdot 3H_2O$	Carbon Co., USA	186 (3.2) 219 (2.4) 405 (3.5)	530 (0.4) 687 (0.3)	376 392	433 (1.6)
Soda, $Na_2CO_3 \cdot 10H_2O$	Synthetic	38 (2.3) 42 (2.1)	118 126 } (4.1)	95 108	92 (1.4)
Trona, $Na_3H(CO_3)_2 \cdot 2H_2O$	Lake Natron, Egypt	117 (0.9)	142 (0.6)	100 126	98 (0.2)
Pirssonite, $Na_2Ca(CO_3)_2 \cdot 2H_2O$	Searles Lake, Calif.	196 (6.5) 268 (0.15) 378 (0.7) 432 (0.5)	806 (1.7) 852 (0.45) 903 (0.15)	182	828 (0.4)
Gaylussite, $Na_2Ca(CO_3)_2 \cdot 5H_2O$	Alerida, Venezuela	57 (0.3) 90 (0.4) 132 (7.3) 176 (0.5) 429 (0.2)	761 (0.3) 806 (1.8) 844 (0.3) 888 (0.2)	111 120	391 (1.4) 823 (0.6)
Gaylussite	Synthetic	61 (0.8) 128 (6.8) 185 (1.0)	803 (3.3) 850 (0.2) 900 (0.3)	723	322 (0.1) 418 (2.8) 815 (1.2)

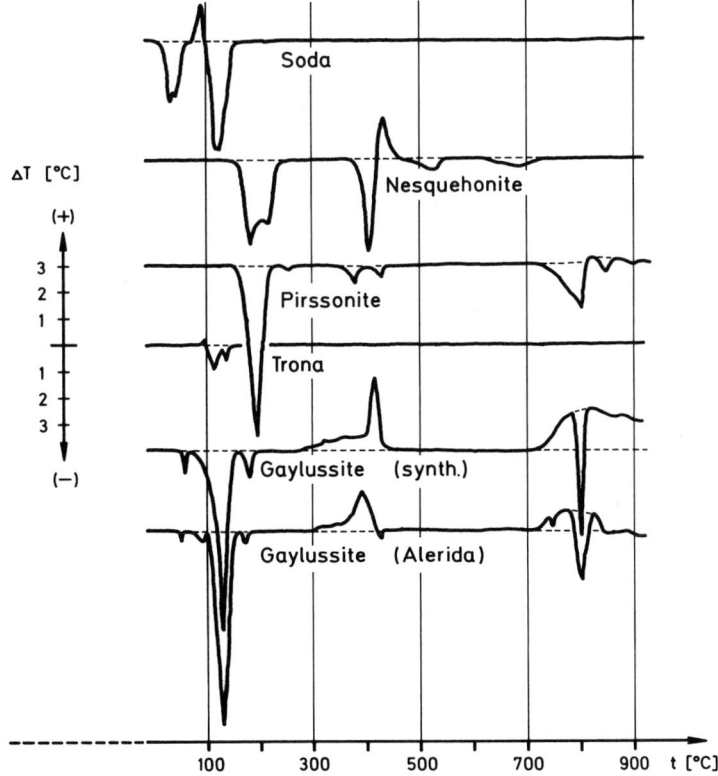

Fig. 22. DTA curves of hydrated carbonates without other anions

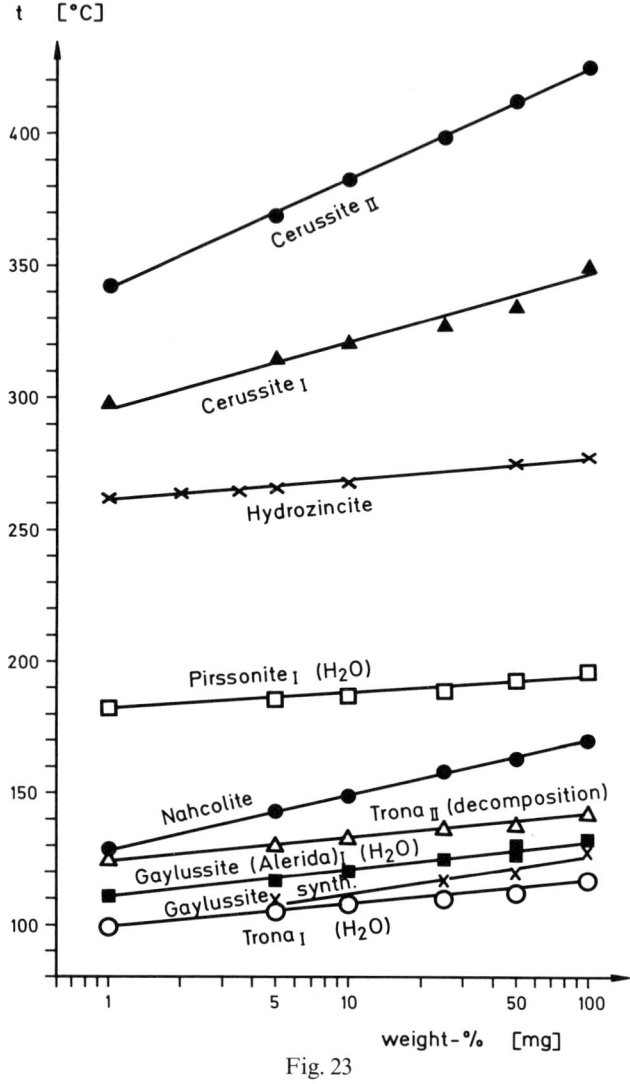

Fig. 23

Figs. 23 and 24. PA-curves of hydrated carbonates without other anions (and of cerussite)

curves of hydrated carbonates differ greatly from those of carbonates free of water by these dehydration peaks.

The Na-Ca-carbonates pirssonite and gaylussite that show very similar DTA curves decompose in two steps. After the decomposition of the Na_2CO_3-part of the structure at $\sim 200°$ C (pirssonite) resp. $\sim 400°$ C

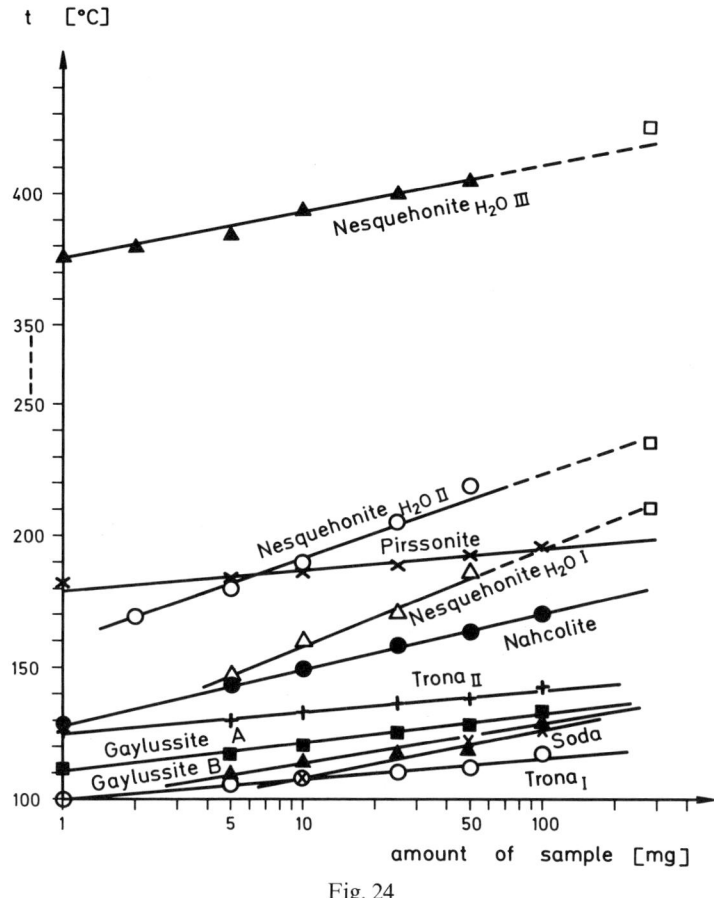

Fig. 24

(gaylussite), $Na_2O + CaCO_3$ still remained. Before this $CaCO_3$ is decomposed, the substance begins to sinter: after 700° C a partial melt rich in sodium will form, as has been observed with other sodium minerals (e.g. in the thermal behaviour of the Na-silicate loughlinite, compare with ECHLE); this process of sintering is reflected by drifting of the zero-line to the endothermic side (see Fig. 22, DTA curves of pirssonite and gaylussite).

4.4 Hydrated Carbonates with Other Anions

The minerals of this group occur very seldom in nature. Some DTA curves of such minerals as schroeckingerite, hydrotalcite or hydromagnesite have been published by BECK (1950). WALENTA published the

Table 13. DTA data of hydrated carbonates with strange anions (in °C)

Mineral (sample location)	Formula	Endothermic reactions (ΔT)			Exothermic reactions (ΔT)
		Dehydration	Decomposition	Stand. temp.	
Hydromagnesite, (Alameda, Calif.)	$Mg_5[(OH)(CO_3)_2] \cdot 4H_2O$	375 (2.5) 422 (1.7) 458 (1.2)	521 (2.6) 566 (2.1)	387 504	501 (5.2)
Artinite (Kraubath, Austria)	$Mg_2[(OH)_2CO_3] \cdot 3H_2O$	270 504	538	240	
Brugnatellite (Valle Malenco, Alps)	$Mg_6Fe^{3+}[(OH)_{13}CO_3] \cdot 4H_2O$	373	447	337 414	
Zaratite (Lancaster, Texas)	$Ni_3[(OH)_4CO_3] \cdot 4H_2O$	280	728	217 426 (!)	
Hydrotalcite (Dypingdalen, Norway)	$Mg_6Al_2[(OH)_{16}CO_3] \cdot 4H_2O$	270	464	210 247	
Schroeckingerite (Sweetwater, Wyo.)	$NaCa_3[F/SO_4/(CO_3)_3] \cdot 10H_2O$ (U-bearing)	174 383	788	108 366 587	
Voglite (Breitenbrunn, Germ.)	Ca-Cu-UO_2-carbonate $\cdot 6H_2O$	80 (0.4) 182 (1.1)	824 (0.2; very broad)	80	712 (0.2)

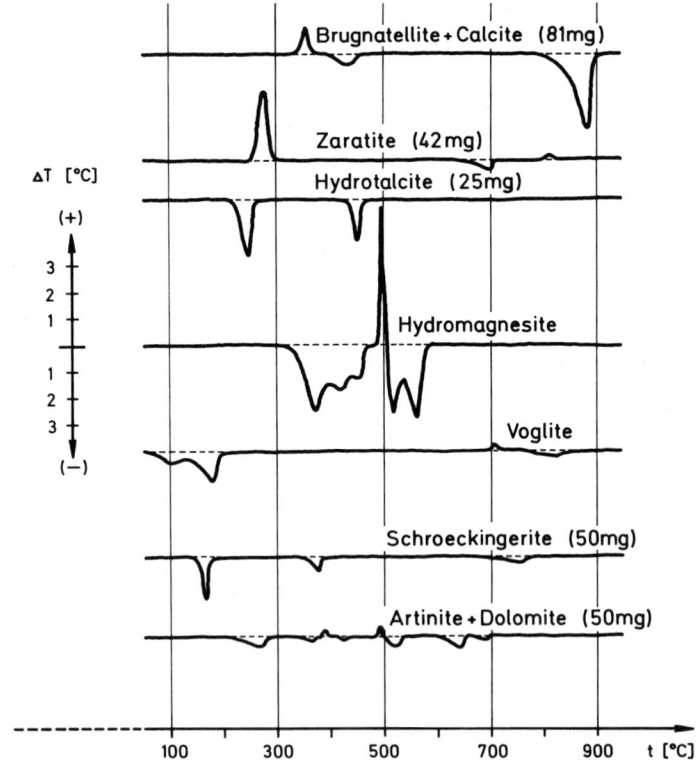

Fig. 25. DTA curves of hydrated carbonates with other anions

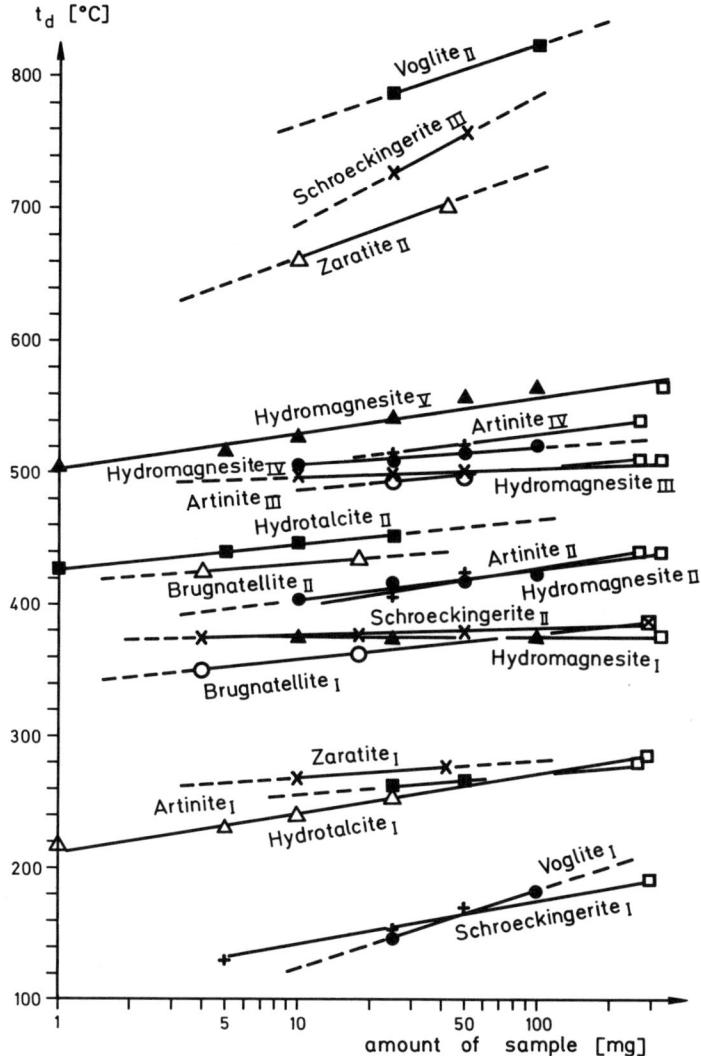

Fig. 26. PA-curves of hydrated carbonates with other anions

DTA data of the Cl-carbonate grimselite. Table 13 contains the DTA data of seven minerals of this group. But since it was not possible to acquire 100 mg amount of sample material of all minerals, the data of Table 13 are partly extrapolated from the PA-curves (Fig. 26). In this case the ΔT-values have not been given. Method of analysis as in 4.1.

All minerals of this group are characterized by minor temperature differences (ΔT) of the decomposition peak: only a small supply of heat

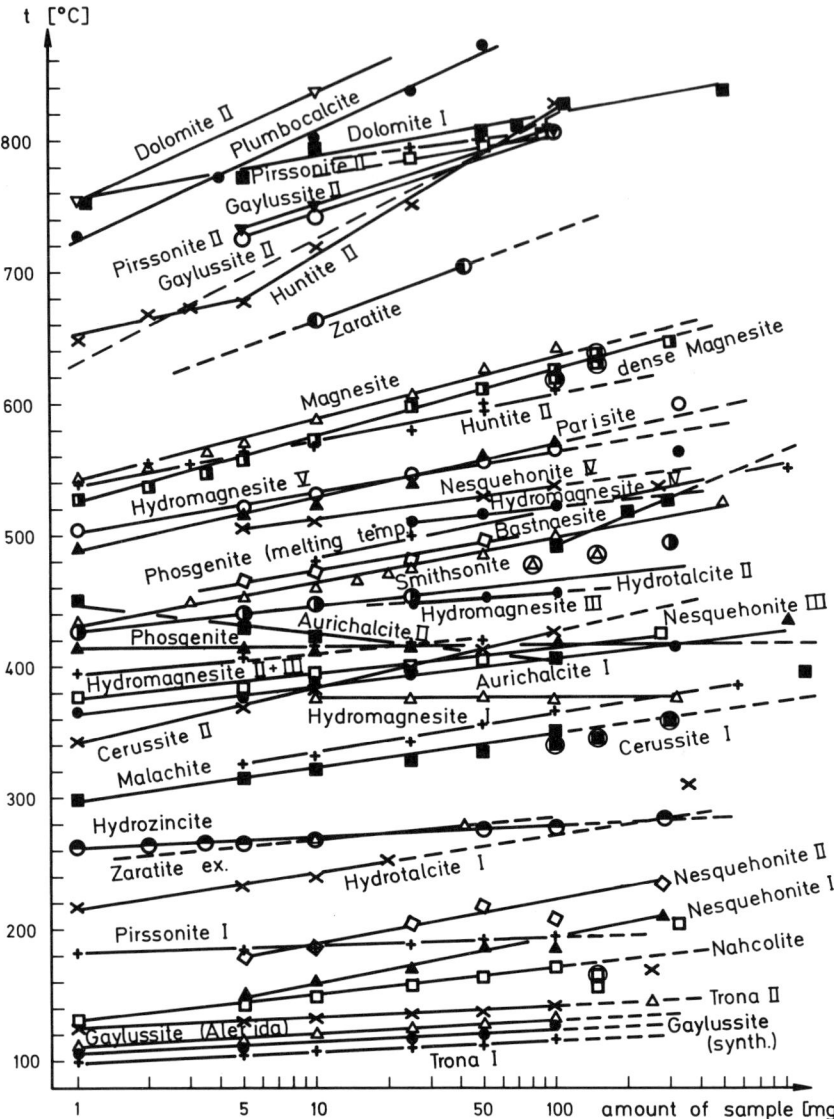

Fig. 27. PA-curves of all carbonates studied in the present investigation. Values for sample amounts of more than 100 mg from BECK (1950)

will be sufficient to destroy the structure of hydrated carbonates with other anions.

Of all carbonate minerals—with the exception of the Fe-Mn-carbonates siderite, ankerite, breunnerite and rhodochrosite—very small amounts of samples can also be proved differential thermal analytically (e.g. 1–3 mg in a 100 mg-sample). So the "standard peak temperatures" (= the beginning of the PA-curves, see Fig. 27) cannot only be obtained by constructing the PA-curves, but also by measurements. For the sure identification of several carbonate minerals occurring in the same rock sample, the heating of different amounts of sample and the drawing of PA-curves will be very useful: values which diverge (differ) from a straight PA-curve in a logarithmic representation will belong to an admixture, and should be identified by comparing with the PA-curves of other carbonates or hydroxides.

4.5 Nitrates

The naturally occurring minerals nitre and Chile nitre can be recognized in DTA by a structural transformation (KNO_3: orthorhombic → trigonal at $128 \pm 0.5°$ C; ΔT: 1.8° for 100 mg) and by their melting temperatures: $NaNO_3$ melts at 314° C $\pm$ 1° (ΔT: 2.2° for 100 mg), KNO_3 at $336 \pm 1°$ C (ΔT: 2.4° C, 100 mg, standard conditions). The transformation temperature of KNO_3 is suitable for calibration as well as for being an internal standard (see III-2.3).

5. Borates, Phosphates, and Arsenates

5.1 Borates

DTA data of borate minerals have been published by BERG (1970: boracite, colemanite, ulexite), HEIDE (1962: ascharite), HURLBUT and ARISTARAIN (1967 a, b); ARISTARAIN and HURLBUT (1967 a, b, 1968: rare borates from deserts in Argentina: rivadavite, ameghinite, ezcurrite, teruggite, macallisterite). Borates containing water resp. OH^- show a dehydration in several steps between 90 and 580° C. Endothermic effects between 600 and 780° C reflect melting of the substances, exothermic effects between 600 and 800° C the reformation of a crystal structure (see Table 14 and Fig. 28).

5.2 Phosphates and Arsenates

Phosphates have to be heated very carefully in DTA, since the molten substances (between 500 and 800° C) can only be removed from the metallic sample holders and from the thermocouples with great difficulty if the phosphates have been in contact with these metals. Solidified melts

Table 14. DTA data of borates, phosphates, and arsenates (in °C)

Mineral (sample location)	Formula	Endothermic reactions (ΔT)		Exothermic reactions (ΔT)
		Dehydration	Decomposition, melting	
Borax (Boron, Calif.)	$Na_2[B_4O_5(OH)_4] \cdot 8H_2O$	100 (0.2) 155 (6.6) 190 (1.5)	630 (0.8)	
Kernite (unknown)	$Na_2[B_4O_6(OH)_2] \cdot 3H_2O$	175 (3.5) 555 (0.3)	755 (0.9)	602 (0.2)
Colemanite (Emet, Turkey)	$Ca[B_2BO_4(OH)_3] \cdot H_2O$	315 (0.2) 405 (8.2)	646 (0.2)	782 (0.6)
Ulexite (unknown)	$NaCa[B_5O_9] \cdot 5H_2O$	145 (4.2) 182 (1.0)		745 (0.4)
Vivianite (Washington D.C.)	$Fe^{2+}[PO_4]_2 \cdot 8H_2O$	122 (0.05) 163 (3.7) 188 (4.3) 199 (4.0)		208 (0.35) 478 (0.2) 633 (0.6) 740 (0.3)
Annabergite (Richelsdorf Mts., Germ.)	$Ni_3[AsO_4]_2 \cdot 8H_2O$	137 (0.2) 266 (8.0)		714 (0.5)
Erythrite (Richelsdorf Mts., Germ.)	$Co_3[AsO_4]_2 \cdot 8H_2O$	152, 235, 301		655 690
Struvite (Dresden, Germ.)	$NH_4Mg[PO_4] \cdot 6H_2O$	116 (0.3) 182 (8.4)		332 (1.0), 465 (0.3), 665 (7.0)
Haidingerite (Joachimsthal, CSSR)	$CaH[AsO_4] \cdot H_2O$	105 (1.0) 135 (2.0) 207 (0.6)	780 (1.0) 820 (0.5) 856 (7.5)	
Amblygonite (Zaire)	$LiAl[(F, OH)PO_4]$		733 (1.5) 802 (4.4)	671 (0.5)

Phosphates and Arsenates

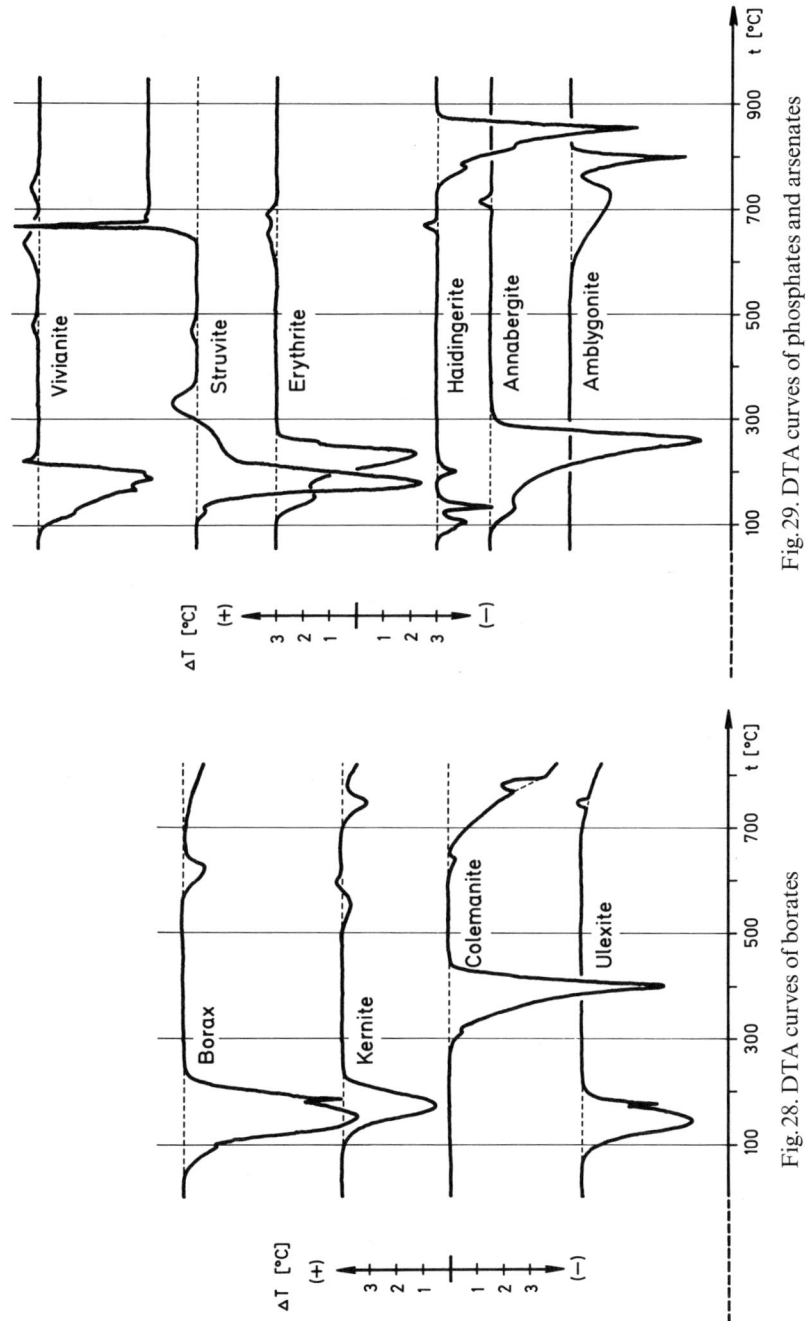

Fig. 29. DTA curves of phosphates and arsenates

Fig. 28. DTA curves of borates

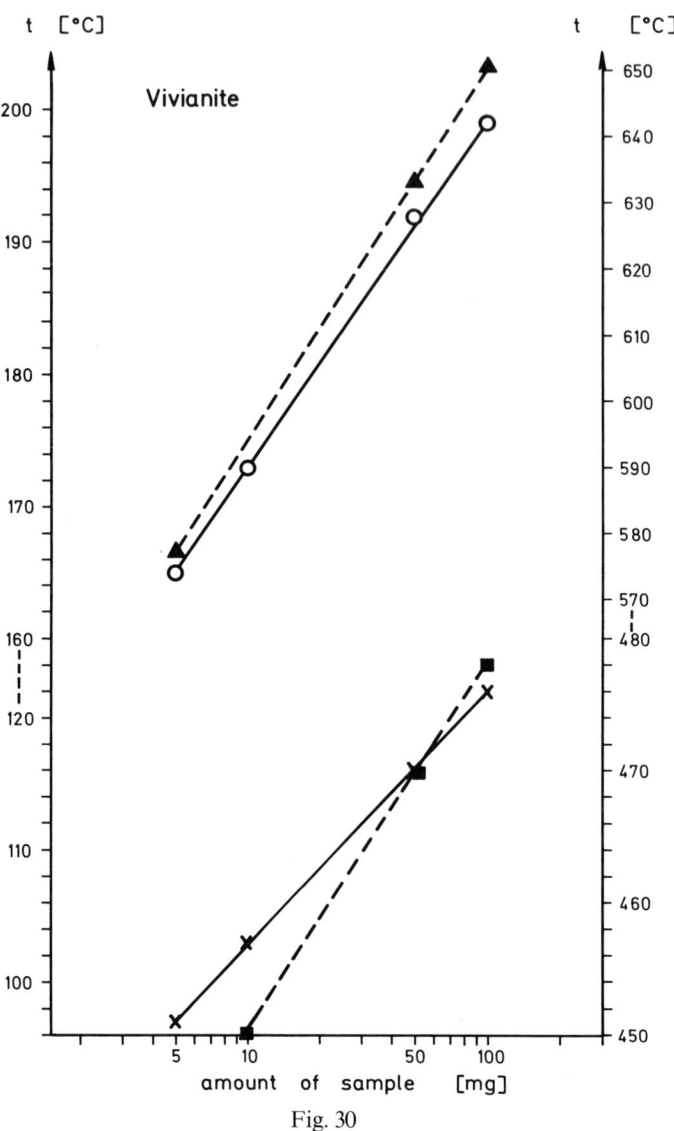

Figs. 30–33. PA-curves of vivianite (Fig. 30), struvite (Fig. 31), annabergite (Fig. 32) and haidingerite (Fig. 33)

of amblygonite are very hard and troublesome. This sticking fast on the metals can be prevented by mixing the phosphate sample with the same amount of Al_2O_3, and by putting a layer of inert material on the thermo-

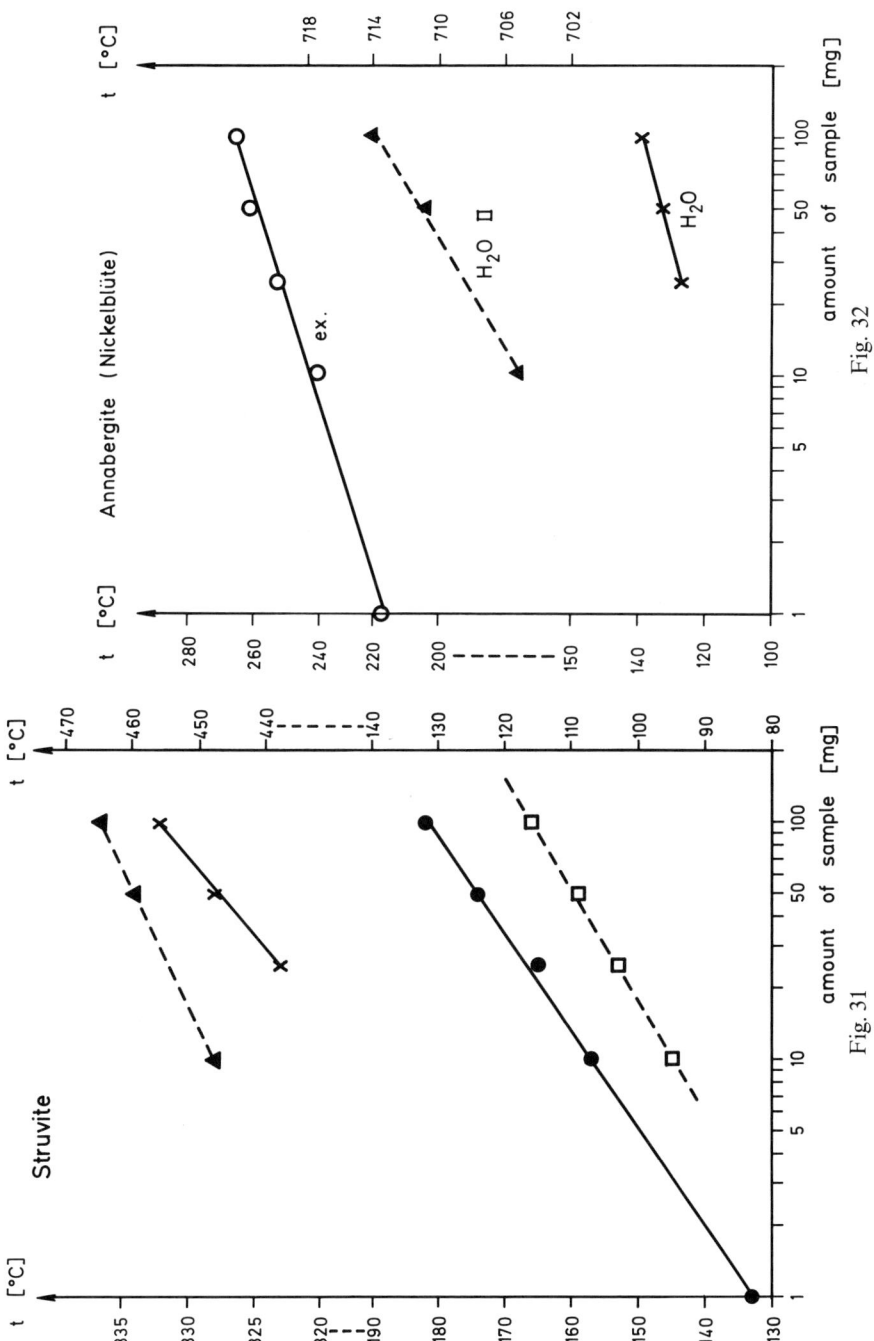

Fig. 31

Fig. 32

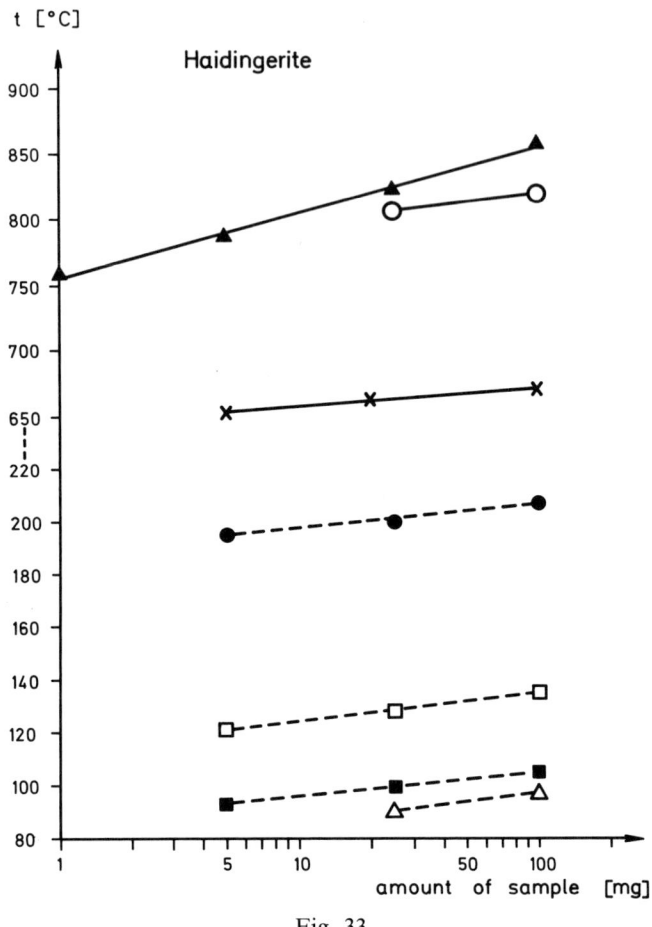

Fig. 33

couples (below the sample), so preventing the sample from contacting the thermocouples.

DTA curves of phosphates (and of the similar minerals of the groups of vanadates or arsenates) are to be found in publications of HAUSEN (schoderite), HURLBUT and ARISTARAIN (1968 a, b: bermanite, beustite), ABELEDO et al. (sanjuanite), CESBRON and FRITSCHE (mounanaite) and VAN WAMBEKE (bolivarite). For the investigation of the minerals listed in Table 14, the standard conditions of analysis have been modified by the use of the Al_2O_3-layer below the sample, by mixing of 100 mg sample amount with 100 mg Al_2O_3, and by the choice of a nickel block for a sample holder. Exactness of measurements: $\pm 1°$ C.

6. Ortho-, Ring-, and Chain Silicates

The silicate minerals summed up in this chapter only show thermal effects below 900° C, measurable in conventional DTA, if they contain water (e.g. the minerals lawsonite, hemimorphite, dioptase, chrysocolla). Generally silicate structures are very stable. So even the dehydration of OH^--containing silicates (with the exception of the sheet silicates) will take place at very high temperatures. This was demonstrated by HUNZIKER when he described the dehydration temperatures of epidotes lying between 960 and 1200° C, or by WITTELS, PETERS (1963), VAN DER PLAS and HÜGI reporting the dehydration of amphiboles lying at temperatures about 1000° C. So do the vesuvianite (PETERS, 1961) and the minerals of the group of tourmalines (see Fig. 34). The DTA characteristics of all these silicates have to be studied in a high-temperature DTA apparatus running up to temperatures of 1500° C. SCHWAB, SCHWAB and JABLONSKI investigated the decomposition and transformation behaviour of some pyroxenes. DIETRICH published the data of the mineral ilvaite, and garnets can also be investigated by means of DTA if they contain some (OH) instead of (SiO_4); e.g. see HEFLIK and ZABINSKI, who reported on the data of hydrogrossular. Thermogravimetric investigations of FREEMAN and of PETERS (1963) lead us to suppose that DTA is a good possibility of classification for amphiboles and epidotes similar to the method described in the present study for chlorites (compare with III-4): FREEMAN found out increasing dehydration temperatures of amphiboles (>980° C) with increasing contents of magnesium. PETERS (1963) observed different DTA curves of epidotes

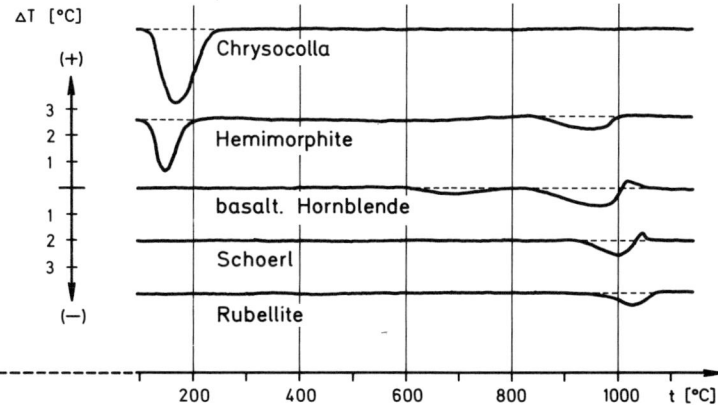

Fig. 34. DTA curves of chrysocolla from Elba, hemimorphite, basaltic hornblende and two tourmalines (schörl from Mull, Scotland; rubellite from Pala, California)

depending on their contents of iron. The DTA curves contained in Fig. 34 were obtained under standard conditions of analysis.

7. Sheet Silicates

Sheet silicates are built up of regular or irregular sequences of tetrahedral $(Si, Al)O_4$-layers, the tetrahedrons of which are connected about three oxygens to a plane network, and of octahedral $(Al, Mg)(OH)_6$-layers, with water and various cations between the different packets. They are generally very fine-grained ($<2\,\mu\varnothing$) and are the "classical terrain" for differential thermal analysis. These silicate minerals occurring in soils and sediments or formed by hydrothermal or metamorphic processes can be identified in DTA curves by their characteristical dehydration behaviour (MACKENZIE, 1970). But a semi-quantitative determination as it has been described for carbonate minerals will be difficult for sheet silicates, especially for the *clay minerals* (kaolinites, montmorillonites, mixed-layers) and for *micas* formed in sediments and built up incompletely at all structural positions (illites, glauconites). This difficulty results from disorder of the structure which lowers the thermal effects, from differences in grain or particle size, from strongly varying chemical compositions of the minerals and from the fact that sheet silicate minerals scarcely occur in a pure form. However, they only occur in mixtures of some minerals which cannot be separated into monomineralic fractions by means of conventional methods of rock fractionation or separation. The determinations of kinetic data of reactions is very inaccurate, even for well-ordered kaolinites: VAN DER MAREL, for instance, obtained for the dehydration of kaolinites heats of reaction between 100 and 176 cal/g. No other group of minerals exists which is characterized by such different DTA data than is the group of vermiculites.

The layer character and the extraordinarily good cleavage after the basis (001) demand for a careful preparation of sheet silicates. MACKENZIE and MILNE did not obtain clear DTA effects of muscovites until after they had ground the samples for some hours. DTA curves of serpentine minerals finally show the dehydration peak after grinding of more than 30 hrs. A lot of DTA data of clay minerals had been compiled by SPEIL et al. (1945), MACKENZIE (1957), BROWN (1961), MCLAUGHLIN (1967), MACKENZIE (1970), BIDLÓ (1971), KURZWEIL (1973) a. o.

7.1 Kaolinites

DTA curves of the minerals of this group show a strong endothermic deflection between 530 and 700° C due to the dehydration and the decomposition of the structure, and a middle-strong exothermic effect between 940 and 1000° C, reflecting the crystallization of a spinel phase (ORCEL; NORTON, 1939a,b; GRIM; HOFMANN and PETERS; PETERS, 1962, 1963; KAUTZ; SMYKATZ-KLOSS, 1966; CAILLÈRE and RODRIGUEZ; LANGER and KERR a. o.). The mineral halloysite, $Al_2[(OH)_4Si_2O_5] \cdot 2H_2O$, is characterized by an additional endothermic peak between 110° and 130° C. The peak is caused by the evaporation of the water which has occupied the interlayer space. (This dehydration of interlayer water of halloysite was not listed in Table 15). In the case of samples containing well-ordered kaolinite, a semi-quantitative determination of this mineral will be possible, if besides the measurement of peak areas (as it has been tried by VAN DER MAREL or SAND and BATES) either the determination of the relation peak area to width at half height (CARTHEW) is made or by

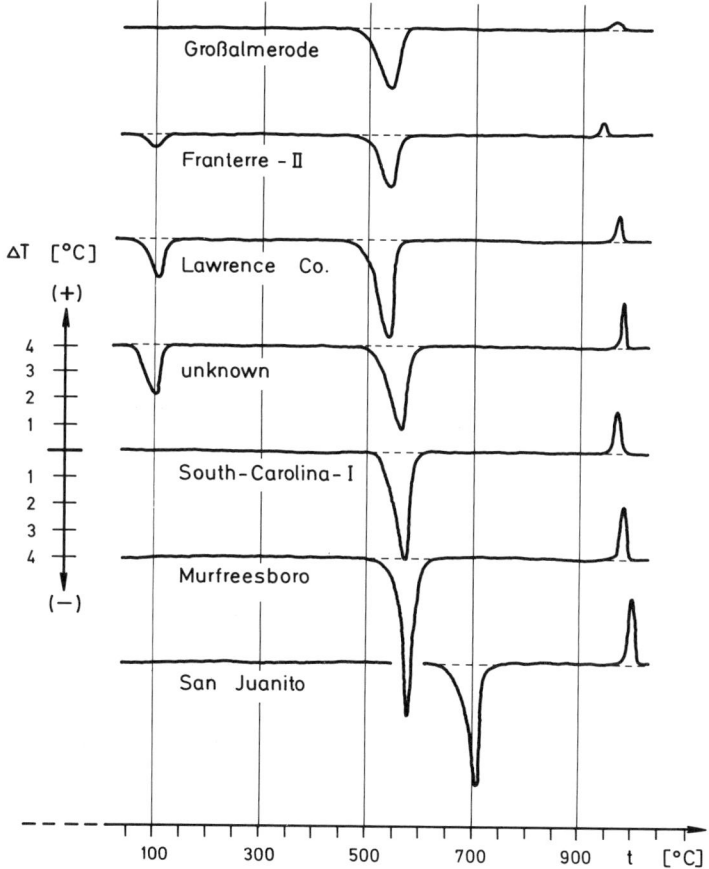

Fig. 35. DTA curves of some minerals of the kaolinite group

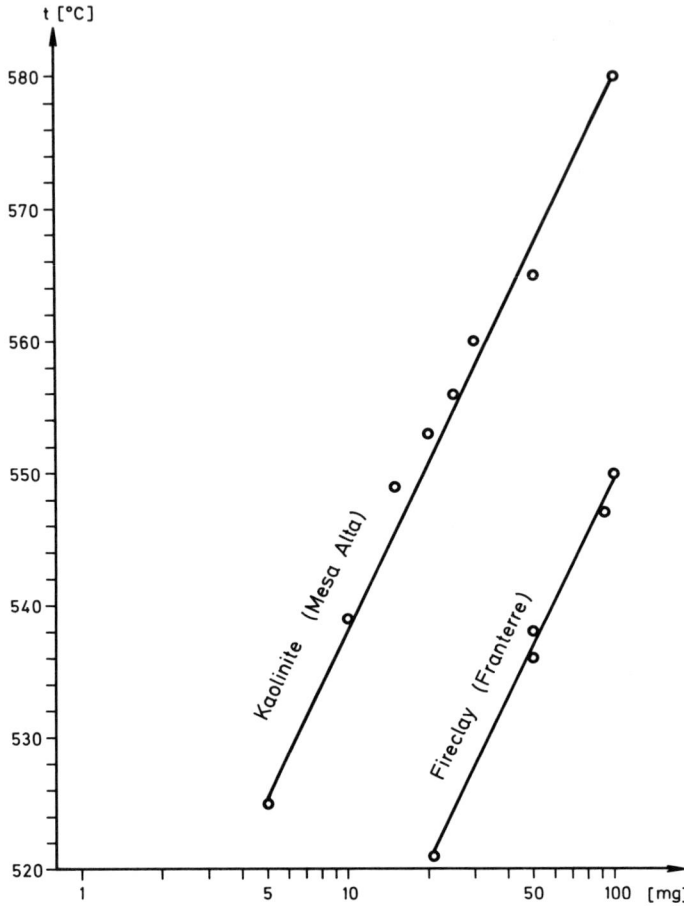

Fig. 36. PA-curves of the well-ordered kaolinite from Mesa Alta and of the disordered fireclay type from Franterre (France)

means of the PA-curve of the endothermic peak (Fig. 36). The dehydration temperatures of kaolinites depend strongly on the degree of disorder of the structure. All change-overs exist between well- and disordered kaolinite (disordered after the crystallographic b-axis), and the DTA is suitable to determine the degree of this disorder (see III-6). The chemical composition of all minerals of this group is $Al_2[(OH)_4Si_2O_5]$, dickite being the hydrothermally formed 2 M-type, fireclay-mineral the variety generally found in soils that is very much disordered in the direction of the b-axis. The data of Table 15 were obtained by standard methods (Table 1). Exactness of measurement: $\pm 1°$ C.

Table 15. DTA data of some kaolin minerals (in °C)

Mineral	Sample locality	Endothermic reaction (ΔT)	Exothermic reaction (ΔT)
Kaolinite	Mesa Alta, N. Mex.	581 (3.9)	1005 (1.3)
Kaolinite	Mesa Alta, N. Mex.	580 (3.2)	983 (2.1)
Kaolinite	Murfreesboro, Arkansas	578 (6.1)	983 (2.0)
Kaolinite	South-Carolina	575 (4.1)	970 (1.5)
Kaolinite	South-Carolina	573 (3.5)	970 (1.6)
Kaolinite	Macon, Georgia	569 (4.0)	994 (1.2)
Halloysite	Unknown	567 (3.3)	983 (1.8)
Halloysite	Djebel Debar, Algerie	555 (6.2)	990 (0.8)
Fireclay-mineral	Franterre, France	543 (2.0)	943 (0.5)
Fireclay-mineral	Franterre, France	550 (3.2)	963 (0.5)
Fireclay-mineral	North-Germany	540 (2.3)	965 (0.3)
Halloysite-clay	Lawrence, Mo.	543 (3.7)	975 (1.0)
Dickite	San Juanito, Mexico	708 (4.6)	999 (2.5)

7.2 Pyrophyllite and Talc

Pyrophyllite and talc being the most simple three-layer silicates, (in which two layers of SiO_4-tetrahedrons include one octahedral layer, the central position of which is occupied by Al^{3+} in the case of pyrophyllite or by Mg^{2+} in the case of talc), they dehydrate and decompose in one combined effect at relatively high temperatures (see Table 16). The slight differences in endothermic peak temperatures of talcs or pyrophyllites from various localities indicate that both minerals are nearly always well-ordered. Contrary to the structure of many kaolinites, montmorillonites or illites, the structures of pyrophyllite and talc are only slightly disordered. Talc has rarely been formed during sedimentary processes (diagenetically in salt rocks), pyrophyllite never: both minerals need

Table 16. DTA data of pyrophyllite and talc (in °C)

Mineral, formula	Sample origin	Endothermic reaction (ΔT)
Pyrophyllite, $Al_2[(OH)_2Si_4O_{10}]$	North-Carolina	745 (2.4)
Talc, $Mg_3[(OH)_2Si_4O_{10}]$	OECD	942 (3.0)
Talc	Greiner, Tyrol	940 (4.6)
Talc (speckstone), white	Göpfersgrün, Bavaria	940 (2.7)
Talc (speckstone), green (impurity of chlorite!)	Göpfersgrün, Bavaria	937 (2.8)

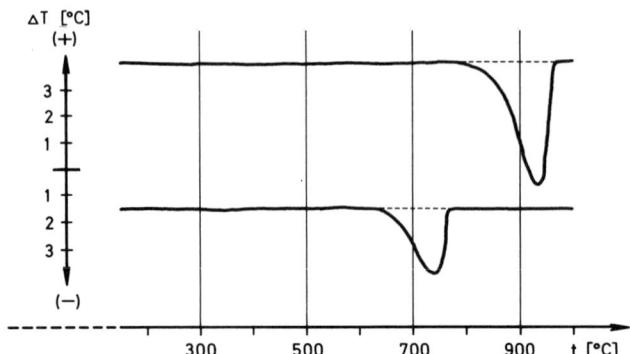

Fig. 37. DTA curves of talc (above) and pyrophyllite (below)

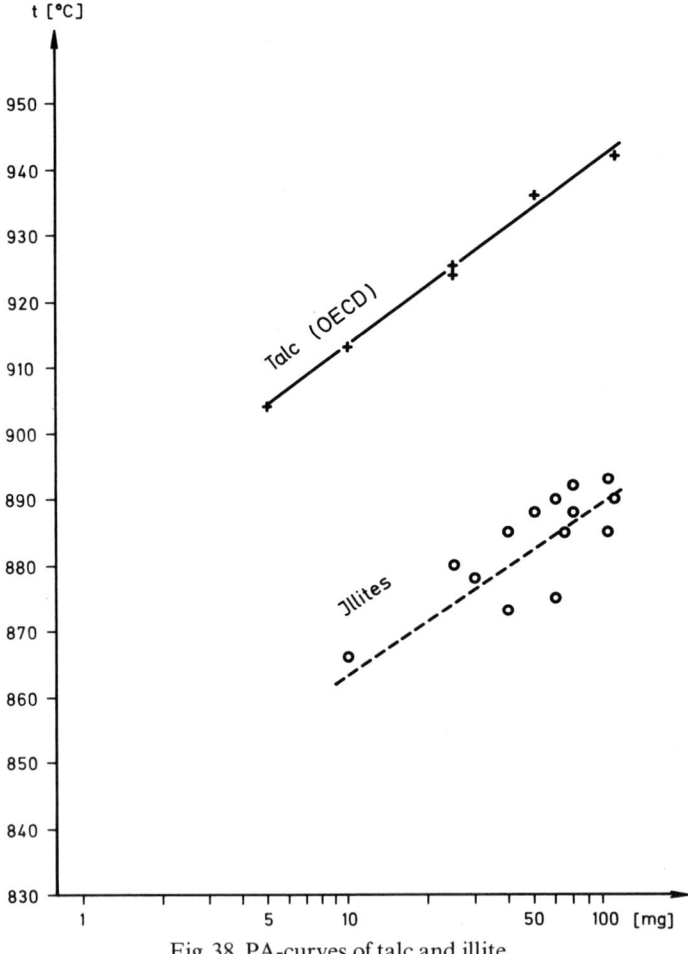

Fig. 38. PA-curves of talc and illite

higher temperatures (and pressures) to be formed, and these higher temperatures are the reason that the minerals are well-ordered. The data of Table 16 were obtained by standard methods of analysis. Exactness of measurements: $\pm 2°$ C.

The data of both the isotype minerals talc and pyrophyllite show great similarity with those of the minerals of the chlorite group (compare with 7.4): talc is dehydrated and decomposed, as well as the pure Mg-chlorite, some hundred degrees later than the corresponding Al-minerals pyrophyllite and Al-chlorite (sudoite). GRIM and ROWLAND (1942) obtained the decomposition peak temperatures of 780° C (pyrophyllite) and of 950° C (talc) for amounts of sample not mentioned.

7.3 Montmorines (Smectites) and Vermiculites

Major differences in the DTA characteristics of montmorines (smectites) and vermiculites have been described (ORCEL and CAILLÈRE; GRIM; GRIM and ROWLAND, 1942, 1944; KULP and KERR, 1948, 1949; BAR-

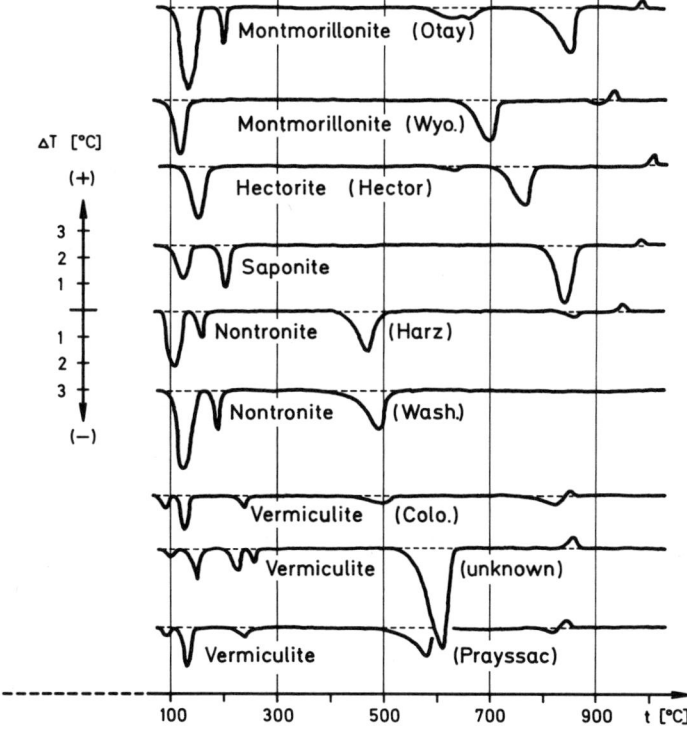

Fig. 39. DTA curves of some montmorillonites and vermiculites

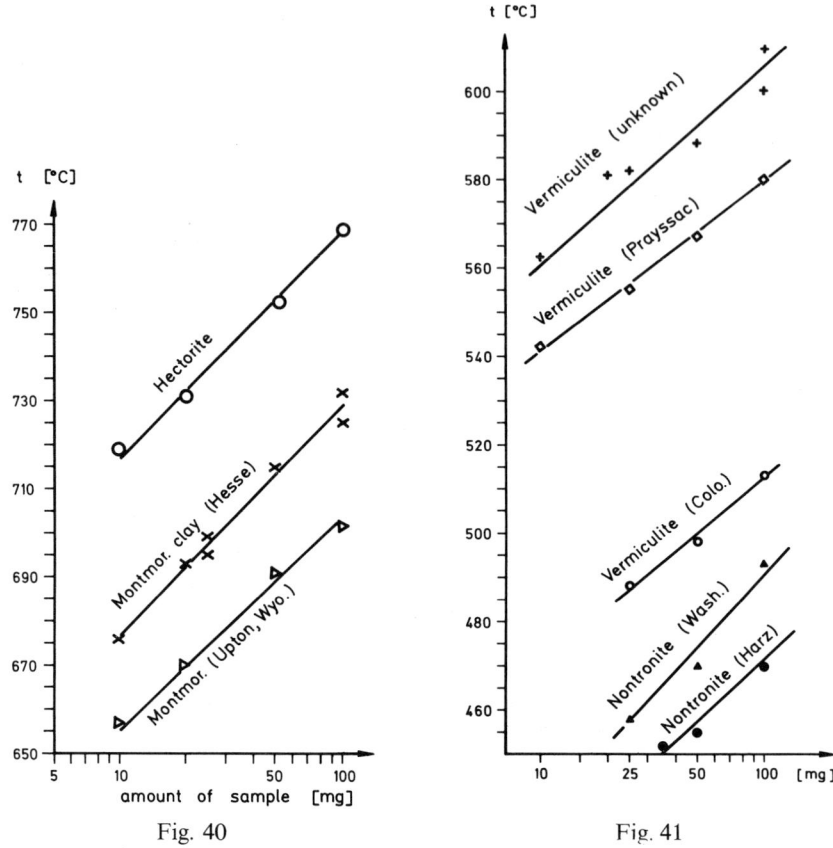

Fig. 40. PA-curves of three montmorillonites

Fig. 41. PA-curves of three vermiculites and of two nontronites (Fe-montmorillonites)

SHAD; FAUST, 1951; EARLEY et al.; WEISS et al.; BASSETT; TAKEUCHI et al.; BOSS; WILSON et al.) due to chemical, crystal physical and grain size differences. The minerals of this group are generally extraordinarily fine-grained, they contain varying amounts of water and cations (Na^+, Ca^{++}, Mg^{++} etc.) between the negatively charged three-layer packets and are frequently clearly disordered. The vermiculites, especially, show such differences in their dehydration and decomposition temperatures (see Fig. 41), that a semi-quantitative determination of vermiculites by means of PA-curves will only be possible if a PA-curve of the one vermiculite of this locality is available. A more detailed differential thermal analytical determination of montmorillonites and vermiculites will of course be possible on account of two characteristic properties of these minerals, the cation exchange capacity and the hydration behaviour (water adsorption capacity, compare with III-5).

Most of the vermiculites are tri-octahedral, that means the central positions in the octahedral layer are occupied nearly to three thirds (= by bivalent ions mainly being Mg^{++}). The dehydration and decomposition temperatures of vermiculites decrease with the increasing portion of trivalent iron (compare with III-5). On the other hand, the frequent smectite minerals are di-octahedral, having occupied the central positions of the octahedrons in the octahedral layer mainly by trivalent ions (then two thirds of these positions are occupied) like Fe^{+++} and Al^{+++}. Here, too, the Fe-mineral nontronite shows the lower dehydration and decomposition temperatures compared to the Al-mineral montmorillonite (see Figs. 40 and 41). The rare tri-octahedral montmorines (hectorite, saponite, sauconite) are decomposed at higher temperatures. The rounded DTA deflections frequently show no distinctly marked top, so the exactness of measurements will be reduced to $\pm 2\text{-}3°$ C for the minerals of this group. The exothermic peak at temperatures of more than 840° C is generally sharp and can be measured exactly ($\pm 0.5°$ C). This peak reflects the recrystallization of a new structure (a kind of mullite), and is the best criterion for the differentiation of vermiculites and smectites appearing in DTA curves of the former between 840 and 890° C, and in curves of the latter at temperatures $>920°$ C. Averaged formulas of the different varieties are (after CORRENS):

Montmorillonite $\{(Al_{1.67}Mg_{0.33})[(OH)_2Si_4O_{10}]^{-0.33} \cdot Na_{0.33}(H_2O)_4\}$

Nontronite $\{Fe_2^{+++}[(OH)_2Al_{0.33}Si_{3.67}O_{10}]^{-0.33} \cdot Na_{0.33}(H_2O)_4\}$

Saponite $\{Mg_3[(OH)_2Al_{0.33}Si_{3.67}O_{10}]^{-0.33} \cdot Na_{0.33}(H_2O)_4\}$

Hectorite $\{$Li- and F-bearing saponite$\}$

Vermiculite $\{Mg_{2.36}Fe_{0.48}^{++}Al_{0.16}^{+}[(OH)_2Al_{1.28}Si_{2.72}O_{10}]^{-0.64} \cdot Mg_{0.32}(H_2O)_4\}$.

7.4 Micas

The preparation of mica samples for DTA investigations has to be made very carefully: only fine-grained micas show clear DTA effects, but sheets of cleavage show none. But grinding in a ball mill (for at least 3 hrs) must not be too strong, since otherwise the structure of the micas will be partly destroyed. Then the decomposition peak will appear 100–200° C earlier (instead of 850° C) in the DTA curves of such partly destroyed micas. The DTA data of the magmatically and metamorphically

Table 17. DTA data of montmorines and vermiculites (in °C)

Mineral, sample locality	Endothermic reactions (ΔT)		Exothermic reactions (ΔT)
	Dehydration (H_2O)	Dehydration (OH) and decomposition	
Montmorillonite, Otay, Calif.	135 (3.0) 197 (1.4)	632 (0.35) 653 (0.35) 855 (1.7)	986 (0.3)
Hectorite, Hector, Calif.	155 (2.0)	634 (0.1) 769 (1.5)	1010 (0.4)
Saponite, Fichtelgeb., Germany	127 (1.3) 204 (1.6)	842 (2.2)	985 (0.2)
Montmorillonite, OECD	132 (2.7) 190 (1.3)	674 (1.2) 880 (0.6)	1000 (0.3)
Montmorillonite, Upton, Wyoming	118 (2.1)	702 (1.5)	936 (0.4)
Montmorillonite-clay, Hesse, Germany	145 (2.5) 220 (0.3)	725 (2.2)	947 (0.2)
Nontronite, St. Andreasberg, Harz/Germany	110 (2.15) 160 (1.6)	470 (1.5) 860 (0.15)	950 (0.25)
Nontronite, Garfield, Washington	126 (3.0) 188 (1.5)	493 (1.5)	
Vermiculite, Colo.	90 (0.5) 125 (1.3)	503 (0.25) 826 (0.35)	851 (0.2)
Vermiculite, Prayssac, France	95 (0.3) 133 (1.5) 240 (0.3)	580 (1.1) 818 (0.15)	845 (0.25)
Vermiculite, unknown	100 (0.3) 151 (1.1) 228 (0.8) 263 (0.5)	610 (3.8)	865 (0.6)
Vermiculite-clay	115 (0.3) 155 (1.0) 271 (0.4)	636 (0.2) 658 (0.2) 851 (0.3)	878 (0.15)

formed micas biotite, muscovite, zinnwaldite, paragonite, phengite, phlogopite have been published by HUNZIKER, MACKENZIE (1970), SCHWANDER et al., MACKENZIE and MILNE, SMYKATZ-KLOSS (1966) and HANSEN. All show the endothermic decomposition and dehydration peak between 850 and 920° C; the data of these authors agree well with the author's investigations (see Table 18). But the data of sedimentary micas and hydromicas differ strongly from data of micas formed at higher temperatures during the magmatic or metamorphic cycle (CUTHBERT;

GRIM and ROWLAND, 1942–1944; GRIM and BRADLEY; WEAVER; KAUTZ a.o.). One reason for the differing thermal behaviour of the sedimentary micas illite and glauconite may be the fact that many of the investigated sedimentary "micas" had been mixed layers of mica and montmorillonite or chlorite. Further differences in chemical composition and grain-size influence the DTA data. The sedimentary micas lose water which has only been bound adsorptively around 100° C, small amounts of water of the interlayer spaces between 120 and 300° C, and OH^- which has been part of the octahedral layers between 500 and 650° C. They are decomposed between 850° and 940° C as in the case of magmatic micas, and begin to sinter after 1100° C. This "sintering peak", which was not listed in Table 17 because it cannot be obtained by a "normal" DTA apparatus, is shifted towards higher temperatures with increasing titanium contents of biotites (HUNZIKER), so being suitable for measuring the Ti contents of biotites if calibration work has to be done. Biotites, moreover, sometimes show a strong exothermic effect between 400 and 600° C, due to the oxidation of bivalent iron (see Fig. 42). The Li-micas lepidolite and zinnwaldite can be recognized by very sharp decomposition peaks with great ΔT (VAN DER MAREL and Table 18). The data of Table 18 were obtained using standard conditions, with the exception of the grinding time (3 hrs!). The PA-curve of an illite is shown in Fig. 38. Exactness of measurements: $\pm 2°$ C.

7.5 Chlorites

The chlorites, three-layer silicates with another octahedral layer ("brucite" or "hydrargillite" layer) between the three-layer packets, vary strongly in their chemical composition. All change-overs exist between the trioctahedral ortho- or talc-chlorites, $\{Mg_3[(OH)_2AlSi_3O_{10}]$, or chamosites, $(Fe^{++}, Fe^{+++})_3 [(OH)_2AlSi_3O_{10}] \cdot (Fe^{++}, Mg)_3(OH,O)_6\}$, and the dioctahedral Al-chlorites in which the trivalent aluminum substitutes the Mg and Fe in the central positions of the octahedral layers. That can be seen very well in DTA curves: DTA is very suitable for the classification of the minerals of the chlorite group (see III-4). Mg-chlorites dehydrate between 600 and 660° C, but will be decomposed 150–250° C later (see Table 19). With increasing Fe-contents the decomposition peak temperature decreases from 860° C (Mg-chlorites free of iron) to $<700°$ C; in pure Fe-chlorites (thuringite, chamosite) the peaks of dehydration and decomposition of the structure coincide, as in the dioctahedral Al-chlorites (sudoite, cookeite). The mineral sudoite shows the lowest decomposition temperature of all chlorites investigated in this study ($<500°$ C!). It is, of course, the only one having been formed

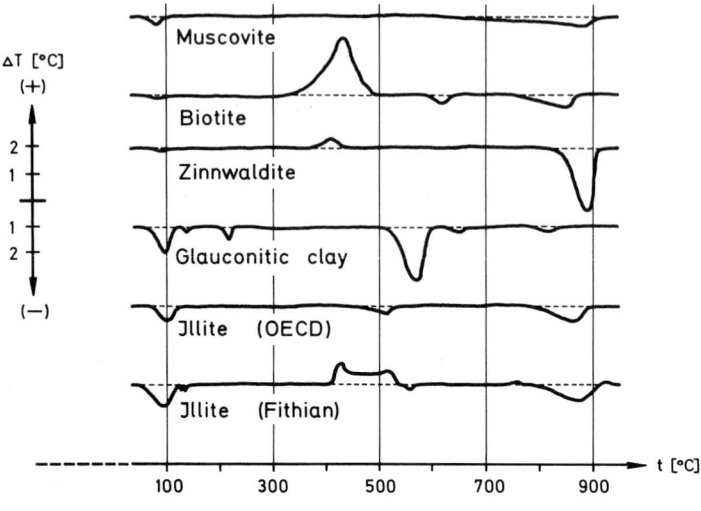

Fig. 42. DTA curves of micas

Table 18. DTA data of micas (in °C)

Mineral, formula	Sample origin	Endothermic reactions (ΔT)			Exothermic reactions (ΔT)
		Dehydration (H_2O)	(OH)	Decomposition	
Sericite (muscovite), $KAl_2[(OH,F)_2AlSi_3O_{10}]$	Valle Antigorio, Ticino, Switzerl.	83 (0.3) broad		887 (0.35)	
Biotite, $K(Mg,Fe,Mn)_3$ $[(OH,F)_2AlSi_3O_{10}]$	Valle Antigorio, Ticino, Switzerl.	90 (0.1)	620 (0.3)	862 (0.4)	435 (2.2)
Zinnwaldite, $K(Li,Fe^{2+},Al)_{2,5-3}$ $[(F,OH)_2Al_{1-0,5}Si_{3-3,5}O_{10}]$	Zinnwald, Erzgebirge, Germany	95 (0.1)		891 (2.5)	415 (0.4)
Illite (sedim. muscovite)	OECD	107 (0.5)	517 (0.3)	864 (0.6)	
Illite-I (0.6–2 µ⌀)	Fithian, Illinois	100 (0.8); 133 (0.2); 142 (0.2)	560 (0.2) 740 (0.15)	870 (0.6)	426 (0.8); 763 (0.1) 510 (0.6); 926 (0.1)
Illite-II (< 0.6 µ⌀)	Fithian, Illinois	98 (0.9); 133 (0.05)	560 (0.4) 740 (0.05)	878 (0.5)	440 (1.0); 763 (0.1) 510 (0.9); 925 (0.1)
Glauconite, (K,Ca,Na) $(Al,Fe^{2+},Fe^{3+},Mg)_2$ $[(OH)_2Si_{3.65}Al_{0.35}O_{10}]$	Uelsen/ Bentheim, Germany	100 (1.0) 142 (0.2) 220 (0.5)		573 (2.1) 658[a] (0.2) 822[a] (0.2)	
Glauconitic sand	Hankenberge/ Osning, Germany	85 (0.5) 195 (0.2) 255 (0.1)	578[a] (0.3)	523 (1.5)	

[a] Impurity of chlorite.

during sedimentary processes (diagenetically in Palaeozoic sandstones). DTA curves of chlorites are to be found in publications of ORCEL, CAILLÈRE and HÉNIN; BRINDLEY; ECKARDT (1967); TROCHIM; ROSS; SCHÜLLER; BORST and KATZ; SUDO et al., ČERNY; WETZEL and others.

Most of the magmatic, metamorphic and sedimentary chlorites are (Mg, Al)-, (Mg, Fe)-, or (Fe, Mg)-chlorites. If Mg and Fe are substituted by Mn^{++} or Ni^{++}, the thermal behaviour should not be modified, since Ni^{++} and Mg^{++} have the same ionic radius and Mn^{++} is only a little larger (for this study no Ni- and Mn-chlorites were available). But the incorporation of Cr^{+++} in the chlorite structure will change the stability of the chlorite structure in a characteristic manner: it can be seen in the DTA curve if the chromium substitutes the Mg or Fe in the octahedral layer, or if it enters the tetrahedral position of silicium. The mineral kaemmererite in which the Cr^{+++} (ionic radius: 0.64 Å) has occupied the position of Mg^{++} (ionic radius: 0.78 Å) in the octahedral layer dehydrates about 100° C later than other chlorites, and the decomposition temperature is also higher, both indicating that the stability of

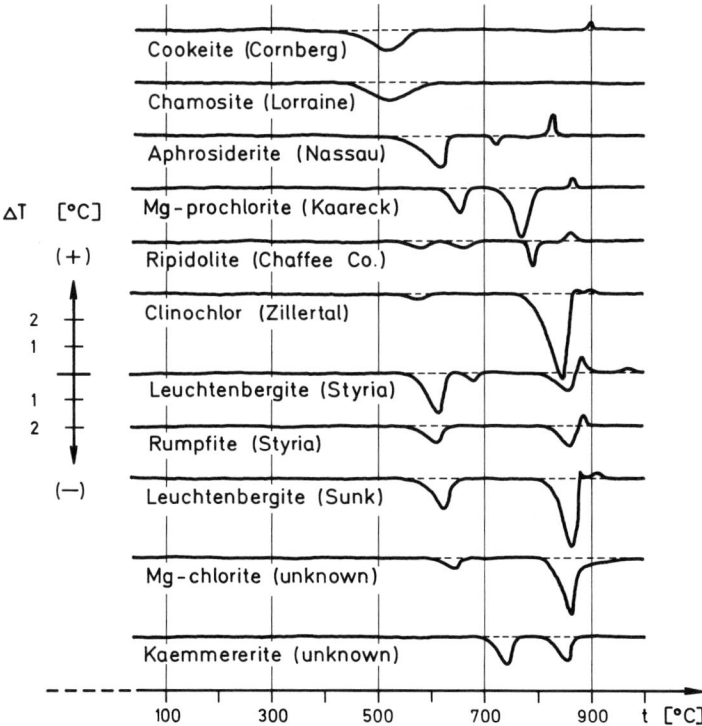

Fig. 43. DTA curves of chlorites. Numbers as in Table 19

Table 19. DTA data of chlorites (in °C)

Mineral	Sample origin	Endothermic reactions (ΔT)		Exothermic reaction (ΔT)
		Dehydrat.	Decompos.	
Leuchtenbergite	Sunk/Trieben, Austria	624 (1.1)	864 (2.6)	878 (0.2) 913 (0.15)
Leuchtenbergite	Kaintaleck, Styria, Austria	613 (1.5) 677 (0.3)	853 (0.7)	881 (0.6) 922 (0.1)
Leuchtenbergite	Ural (USSR)	628 (1.0)	837 (0.5)	905 (0.1)
Mg-chlorite	Neuberg, Austria	630 (5.5)	845 (0.6)	874 (1.8) 925 (0.1)
Ripidolite	Chaffee Co., USA	581 (0.2) 650 (0.2)	791 (1.0)	860 (0.3)
Clinochlor	Zillertal, Austria	580 (0.1)	847 (3.2)	872 (0.1) 900 (0.05)
Mg-prochlorite	Kaareck, Austria	656 (0.8)	773 (1.9)	865 (0.3)
Aphrosiderite	Nassau, Germany	618 (1.2)	721 (0.3)	828 (0.8)
Kaemmererite ($Cr^{[6]}$)		744 (1.0)	859 (0.9)	
Kotschubeite ($Cr^{[4]}$)		604 (0.4) 639 (0.6)	801 (0.2)	828 (0.2)
Pseudo-thuringite	Carinthia, Austria		607 (2.2)	828 (1.2)
Fe-chlorite	Canaglia, Alps		541 (2.0)	
Fe-chlorite (Lias)	CSSR		526 (0.6)	
Thuringite	Schmiedefeld, Sax.		539 (1.9)	
Chamosite (Minette)	Lorraine, France		523 (0.6)	
Al-chlorite	Cornberg, Hesse, Germany		498 (0.8)	898 (0.3)
Prochlorite	Marktredwitz, Bavaria	614 (2.1)	827 (0.5)	848 (0.8)
Prochlorite	Marktredwitz, Bavaria, $> 6\,\mu\emptyset$	617 (2.2)	833 (0.9)	855 (0.8)
Prochlorite	Marktredwitz, Bavaria, $< 6\,\mu\emptyset$	614 (2.2)	833 (1.2)	853 (1.0)
Prochlorite	Marktredwitz, Bavaria, $> 2\,\mu\emptyset$	610 (1.7)	833 (0.8)	850 (0.7)
Prochlorite	Marktredwitz, Bavaria, $< 2\,\mu\emptyset$	616 (1.75)	827 (0.5)	847 (0.7)

the structure has been increased by incorporation of Cr^{+++} in the octahedral layer. On the other hand, the stability of the chlorite structure will be decreased if Cr^{+++} has been incorporated in the tetrahedral layer as a substitute for the Al^{+++} (ionic radius: 0.57 Å). This decreased structure stability is reflected in lower peak temperatures of the second endothermic DTA deflection of kotschubeites (see Table 19). The temperature of the first exothermic peak, which points out the formation of olivine and spinel after the decomposition of the chlorite structure, as well as the decomposition temperature, depends on the chemical composition of the chlorite (compare with III-4). Pure Fe-chlorites have no exothermic DTA peak (see Fig.43 and Table 19), in curves of Fe-Mg-

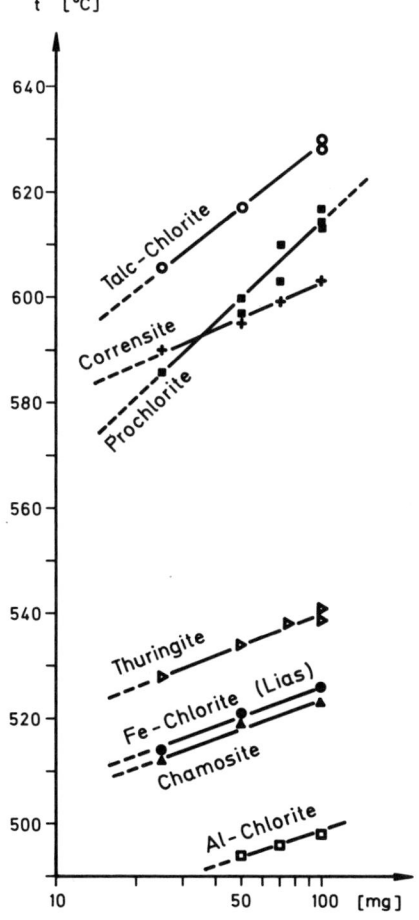

Fig. 44. PA-curves of some chlorites and of corrensite

chlorites (aphrosiderites, pseudothuringites) the exothermic peak appears at 828° C, in curves of Mg-Fe-chlorites (ripidolite, clinochlor) between 845 and 872° C, and in curves of pure Mg-chlorites around 880° C. The highest temperature of this exothermic peak is shown by the Al-chlorite (sudoite).

There are no remarkable differences to bee seen in the DTA curves of the several grain-size fractions of a prochlorite (the last five samples of Table 19). The data of table 19 were obtained using standard conditions, with the exception of the sample holder (nickel block used instead of Pt crucibles), and of the preparation (grinding for 3 hrs in an agate ball mill). Exactness of measurement: $\pm 1°$ C.

7.6 Serpentines

The serpentine minerals are structurally the tri-octahedral analoga to the di-octahedral kaolinites (CORRENS), e.g. the minerals antigorite (sheet serpentine), chrysotile (fibrous serpentine) and lizardite, all being $(Mg, Fe^{2+})_6 [(OH)_8 Si_4 O_{10}]$. Yet the thermal behaviour is very similar to that of the chlorites, as has been shown by WONDRATSCHEK; MIDGLEY; PETERS (1962); HAYASHI, KORSHI and SAKABE; SAITO et al.; NAUMANN and DRESHER; PUSZTASZERI; BASTA and KADER; DIETRICH and above all by CAILLÈRE (1936). The different varieties of serpentine do not differ strongly in DTA (e.g. see Fig. 45). Though the DTA curves of serpentines and chlorites are similar, the first show decomposition temperatures that are 50–80° C lower than those of chlorites, and the temperature of the exothermic peak is also somewhat lower in the case of serpentine minerals. However, as in DTA curves of chlorites, there is the

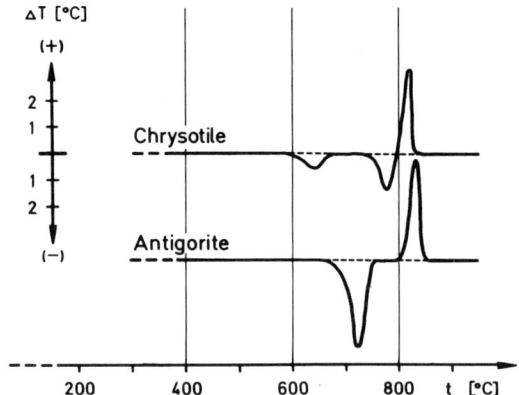

Fig. 45. DTA curves of chrysotile and antigorite

Table 20. DTA data of serpentines (in °C)

Mineral, sample location	Endothermic reactions (ΔT)		Exothermic reaction (ΔT)
	Dehydration	Decomposition	
Chrysotile, Lake Marmorera (Graubünden, Switzerland)	640 (0.5)	774 (1.3)	822 (3.2)
Antigorite, Valle Antigorio (Alps, Switzerland)		722 (2.8)	835 (4.4)

possibility for serpentines, too, to conclude from decomposition temperatures on the Fe-contents: serpentines free of Fe were decomposed between 760 and 800° C, varieties rich in iron were decomposed between 600 and 650° C (quantitative determinations to this point have been started). The exothermic peak between 800 and 835° C is frequently very strong (Fig. 45). Data of Table 20 were obtained from two samples which had been ground in an agate ball mill for 25 hrs (X-ray measurements showed that the crystal structure of the minerals had not been influenced greatly by this long time of grinding), and had then been investigated by DTA under standard conditions. Exactness of measurements: ±0.5° C.

7.7 Palygorskite and Sepiolite

DTA data on palygorskite (attapulgite), $(Mg, Al)_2[OH\,Si_4O_{10}] \cdot 2H_2O + 2H_2O$, and sepiolite ("Meerschaum"), $(Mg, Fe)_4[(OH)_2Si_6O_{15}] \cdot 2H_2O + 4H_2O$, have been published by KULP and KERR (1949); SIDDIQUI; IMAI et al.: HAYASHI et al.; BIDLÓ, VIVALDI and HACH-ALI; MÜLLER-VONMOOS and SCHINDLER. The data vary with different chemical composition: the endothermic main effect (dehydration and partly decomposition) is lowered in its peak temperature by Fe incorporated in the sepiolite structure resp. by incorporation of Fe+Al in the palygorskite structure substituting the Mg in both cases (compare with Figs. 46 and 47). Both minerals contain two kinds of water, the one being in structure channels like in zeolites and being dehydrated between 100 and 130° C, the other (the constitution water) being dehydrated < 320° C (in palygorskites) resp. < 400° C (in sepiolites).

In DTA curves of sepiolites, the exothermic peak reflecting the formation of olivine and spinel appears nearly 50° C earlier than it does in the curves of palygorskites. Data of Table 21 using standard conditions. Exactness of measurements: ±1° C.

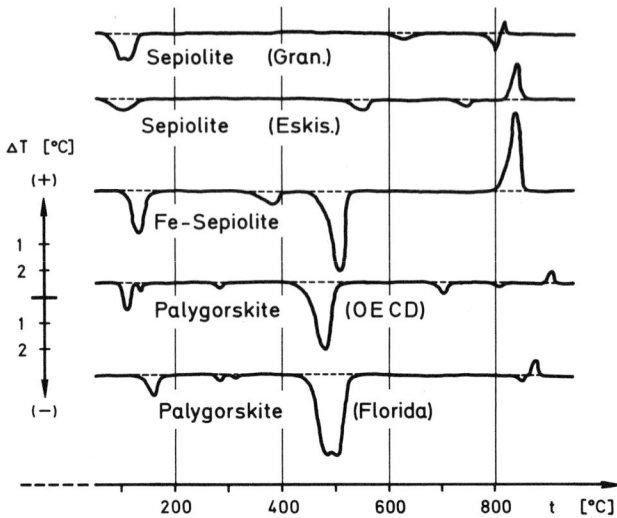

Fig. 46. DTA curves of palygorskites and sepiolites

Table 21. DTA data of palygorskites and sepiolites (in °C)

Mineral, sample origin	Endothermic reactions (ΔT)		Exothermic reactions (ΔT)
	Dehydration (H_2O)	OH-release and decomposition	
Palygorskite, OECD	108 (1.0) 139 (0.3) 285 (0.2)	478 (2.5) 706 (0.4) 808 (0.1)	910 (0.5)
Palygorskite, Midway Mine, Florida	160 (0.8) 285 (0.2) 316 (0.1)	482 ⎫ 509 ⎬ (3.0) 853 (0.2)	880 (0.6)
Sepiolite, Granada, Spain	99 ⎫ 113 ⎭ (1.0)	630 (0.2) 800 (0.6)	824 (0.4)
Sepiolite Eskišehir, Turkey	100 (0.4; broad)	550 (0.4) 745 (0.2)	840 (1.0)
Sepiolite (Meerschaum), Eskišehir	80 (0.2; broad)	577 (0.4) 756 (0.25)	835 (0.8)
Fe-sepiolite, Eskišehir	128 (1.6) 387 (0.5)	506 (3.0)	840 (3.0)

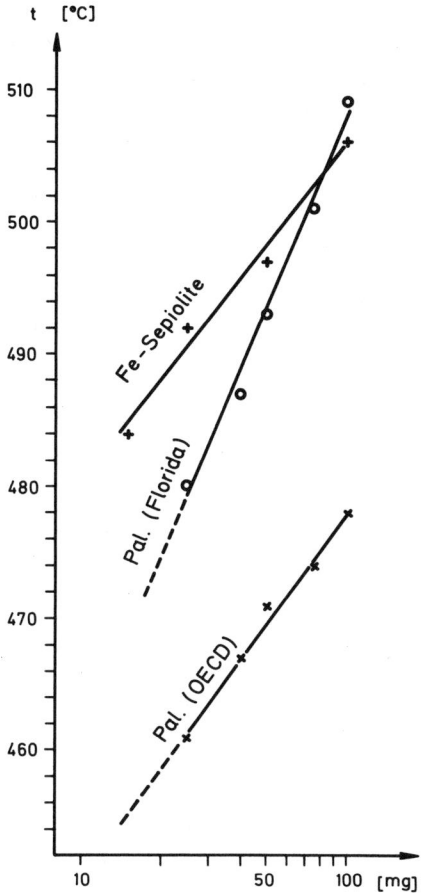

Fig. 47. PA-curves of Fe-sepiolite and of two palygorskites

7.8 Clay Minerals with Mixed-Layer Structure

The properties of clay minerals with mixed-layer structure ("mixed layers") are not necessarily the same as those of their components (SUDO and SHIMODA), so that in thermal behaviour the mixed layers also differ from that of their components. Only very few DTA data on this group of minerals exist, and these data mostly relate to the regular mixed-layer structures which could be identified much more easily than the numerously occurring irregular mixed layers. Regular types are rectorite (mica-montmorillonite mixed layer, KODAMA), tosudite (Al-chlorite-montmorillonite mixed layer, SUDO et al.) and corrensite (swelling chlorite-chlorite mixed layer, LIPPMANN, 1954, 1959). Irregular mixed layers belong to the most frequent minerals of soils and young sediments which can generally be determined only by means of special investigation methods like combined hydration, X-ray tectural and heating techniques (BROWN). Very often the minerals of this type have been neglected in sedimentologic or soil scientific work. By means of DTA up to now only irregular mixed layers of illite and montmorillonite have been investigated (COLE, BALL): COLE distinguished two series of this mineral type depending on the temperature of the main endothermic peak reflecting the dehydration. They differ according to whether this peak appears close to 500° C (predominance of the illitic component), or close to 700° C (predominance of the montmorillonitic component). More suitable for the estimation of the portions of the components than COLE's method will be the measurement of the temperature interval between the two endothermic peaks near 500 and 700° C, and the comparison of this temperature interval with that of the pure component minerals of the same grain-size. Table 22 contains the data of corrensite, of four irregular chlorite-montmorillonite and of one irregular illite-montmoril-

Table 22. DTA data of mixed-layer minerals (in °C)

Mineral, sample origin	Endothermic reactions (ΔT)			Exothermic reactions (ΔT)	
	Dehydration		Decomposition		a-value
	H_2O	OH			
Corrensite, North Hesse	110 (0.4)	600 (1.5)	825 (1.0)	845 (0.7)	20
Corrensite, North Hesse, $>2\mu\varnothing$	95 (0.6)	603 (1.9)	827 (1.0)	848 (0.7)	21
Irregular chlorite-montmorillonite mixed-layer, Watt of the Flensburger Förde, Germany	117, 138 (0.6) (0.2)	640, 675 (0.7) (0.5)	815 (0.1)	865 (0.2)	50
Irregular chlorite-montmorillonite mixed-layer, Scharzfeld/Harz, Germany	115, 146 (0.7) (0.2)	635, 665 (0.7) (0.3)	830 (0.2)	875 (0.2)	45
Dito, second sample	112, 143 (0.5) (0.4)	612, 686 (0.3) (0.5)	770 (0.2)	800 (0.2)	30
Dito, Bartolfelde/Harz, Germany	115, 134 (0.6) (0.2)	632, 660 (0.6) (0.3)	796 (0.2)	830 (0.2)	34
Irregular illite-montmorillonite mixed-layer, North Hesse	98, 202 (0.4) (0.25)	533, 557 (1.3), 680 (0.4)	763 (0.05)	808, 897 (0.05) (0.1)	45

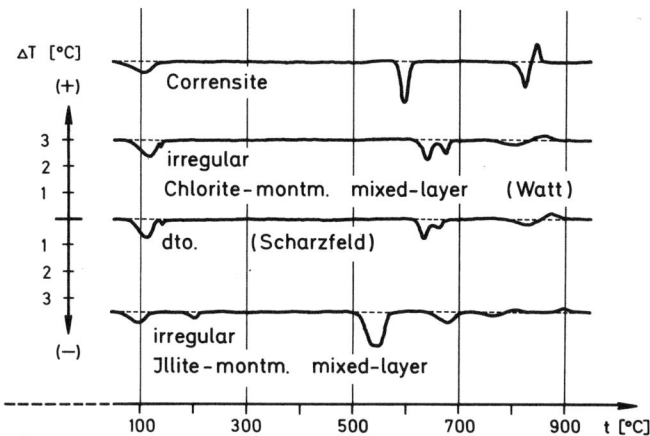

Fig. 48. DTA curves of some clay minerals with mixed-layer structures

lonite mixed-layer minerals, all being investigated under standard conditions with the exception of the sample holder (in this case: nickel block). Exactness of measurements: $\pm 1°$ C.

The DTA data of mixed layers are indeed similar to those of their components. But the peak temperatures lie [analogous to the (001) X-ray interferences] *between* those of their components (with the exception of the dehydration temperatures $<250°$ C). For example, in the DTA curve of an irregular chlorite-montmorillonite mixed layer the dehydration of the "chlorite"-part will appear 20–40° C higher, the adequate dehydration of "montmorillonite" 40–20° C lower than in prepared mixtures of the pure components (SMYKATZ-KLOSS, 1966). For recognition of a mixed layer the PA-curves of the pure components are very useful.

Many sheet silicates rebuild a new structure after being decomposed as can be seen by the exothermic DTA effect between 750 and 1000° C. The temperature of this exothermic effect is very characteristic for the chemical composition of the heated substance. The *temperature interval* between this exothermic peak and the endothermic peak which has appeared immediately before the exothermic effect, reflecting the complete decomposition of the original structure, can be taken as a *measure for the regularity resp. irregularity of the layer sequences* if minerals are low in iron contents and if the grain size of minerals are comparable. The data of nearly 60 investigated sheet silicate samples show temperature intervals that are very specific for the types of minerals (see Table 23); so the *new DTA characteristic a* (= the temperature interval between the last decomposition peak and the following exothermic peak) can be proposed for clay minerals.

Table 23. a-Values of sheet silicate minerals

Type of mineral	Amount of sample (number of samples)	a-Values (° C) minimal	a-Values (° C) maximal	a-Values (°C), average value
Sepiolites	2	20	20	20
Mg-rich chlorites	12	14	29	22
Corrensites	5	19	25	22
Vermiculites	8	20	27	24
Palygorskites	3	26	27	27
Montmorillonites	7	25	43	32
Irregular chlorite-montmorillonite mixed-layers	4	30	50	39
Irregular illite-montmorillonite mixed-layers	3	38	52	44
Chrysotiles (fibrous serp.)	1		48	48
Antigorites	1		113	113
Fe-rich minerals:				
Fe-sepiolites	3	79	95	86
Fe-chlorites	3	69	107	90
Fe-Al-rich palygorskites	1		102	102
Nontronites (Fe-montmor.)	1		81	81

Striking are the very high a-values of minerals rich in iron, reflecting the fact that iron-rich clay minerals are mostly decomposed very much earlier than analogous Mg-silicates of the same structure, and the great a-values of serpentines. But the serpentines especially show a strongly deformed habit, the chrysotiles being rolled-up, the antigorites having a wavy arrangement of the sheet packets and a large degree of disorder (CORRENS). While the regular mixed layer corrensite only shows a minor a-value like its components, the irregular mixed layers always have higher a-values than their components. Besides that, the irregular mixed layers show DTA deflections being lower in ΔT and more rounded than those of their components (compare with Fig. 48). More detailed investigations on the dependence of a-values on grain size and chemical composition and therefore about the reliability of these a-values for measurements of structural disorder, especially as a measure of the irregularity of mixed layers, have been started.

7.9 Mixtures of Sedimentary Minerals ("Clays")

Besides mixed layers the frequently occurring minerals of soils and pelitic sediments are illite, chlorite, kaolinite, montmorillonite, goethite

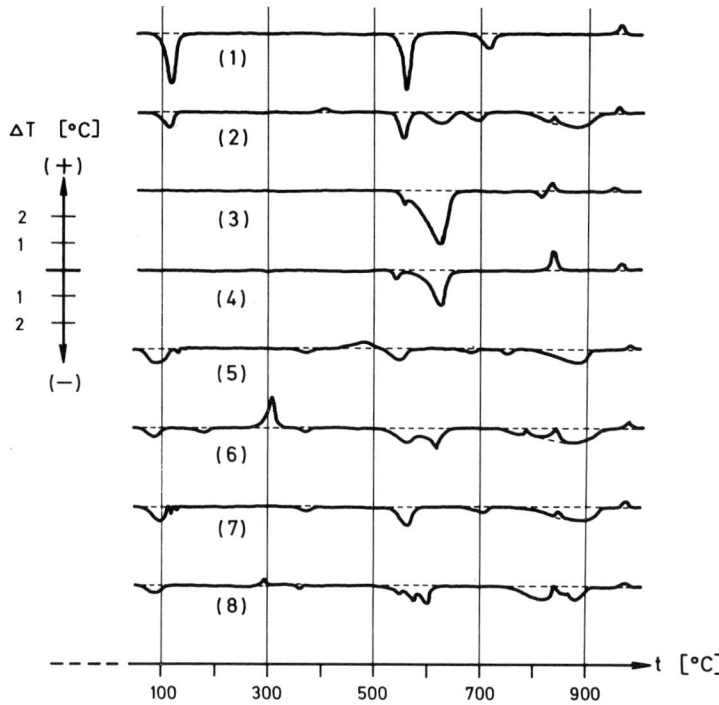

Fig. 49. DTA curves of mixtures of sedimentary minerals

and quartz. Therefore 15 mixtures of these minerals (grain size fraction $<2\,\mu\varnothing$) were heated differential thermal analytically under standard conditions (but with a nickel block as sample holder). The data (Table 24) allow the following statements:

1. Qualitative DTA determinations of pelitic rocks are possible down to 5% of a sheet silicate mineral in a 100 mg-sample, 3 mg of goethite and 15 mg of quartz. All DTA effects can be identified as belonging to a distinct mineral, with the exception of the endothermic deflection at $\sim 550°$ C caused by the release of OH^- from a clay mineral structure; this deflection may belong either to illite or kaolinite or to a Fe-rich chlorite.

2. Semi-quantitative determinations are only possible in the case of minerals with thermal effects not lying too close together. That will be possible in mixtures of kaolinite and montmorillonite, montmorillonite and chlorite or illite, illite and chlorite (+ goethite in all cases). The DTA curves of some of such mixtures are figured above (Fig. 49).

Table 24. DTA of mixtures of sedimentary

Minerals (mg) k = kaolinite, q = quartz, g = goethite, i = illite, ch = chlorite, m = montmorillonite	Endothermic reactions in °C (ΔT, °C) Dehydration				
	H$_2$O		OH		
	adsorpt.	Interlayer	g	k+i	ch
k (50), m (50)		117 (2.0)		562 (2.2)	
ch (95), k (5)				555 (0.1)	623 (2.0)
ch (90), k (10)				542 (0.2)	627 (1.3)
ch (85), k (15)				550 (0.5)	625 (1.5)
i (95), k (5)		118, 135, 165 (0.25) (0.2) (0.2)		530 (0.05) 550 (0.7)	
i (70), k (30)	100 (0.3)	130, 142 (0.1) (0.1)		560 (1.6)	
i (25), k (25), m (25), ch (25)		116 (0.6)		556 (1.0)	630 (0.4)
q (3), g (7), k (20), i (60), m (10)	93 (0.5)	135 (0.05)	373 (0.1)	548 (0.35)	
q (3), g (7), k (10), i (70), m (10)		105, 133 (0.5) (0.03)	370 (0.1)	552 (0.3)	
q (3), g (7), k (10), i (60), m (20)	100 (0.7)	118, 137 (0.25) (0.1)	370 (0.1)	547 (0.4)	
q (3), g (7), k (28), i (40), m (2), ch (20)	87 (0.3)	173 (0.15)	370 (0.1)	563 (0.5)	615 (0.6)
q (3), g (7), k (30), i (40), m (20)	100 (0.5)	118, 137 (0.2) (0.05)	371 (0.1)	563 (0.7)	
q (3), g (7), k (20), i (65), m (5)	100 (0.5)		368 (0.1)	568 (0.6)	
q (51), g (3), k (12.5), i (17.5), m (1), ch (15)	90 (0.3)		361 (0.03)	543 (0.2)	600 (0.4)
q (51), g (3.5), k (5), i (33.5), m (1.5), ch (5)	90 (0.3)		361 (0.05)	540 (0.05)	596 (0.15)

minerals (sample amounts: 100 mg)

							Exothermic reactions in °C (ΔT, °C)	
Inversion	Decomposition							
q	m	i	ch+m	i	i	i	ch+m	k
	715 (0.5)							963 (0.3)
			817 (0.2)				835 (0.3)	950 (0.05)
							834 (0.7)	967 (0.2)
			809 (0.1)				833 (0.4)	963 (0.3)
				885 (2.5)	433 (0.4)	505 (0.3)		950 (0.05)
				888 (1.0)	280 (0.15)	415 465 (0.2)		968 (0.5)
	695 (0.3)			880 (0.4)	405 (0.1)		836 (0.2)	963 (0.2)
	687 (0.05)	753 (0.1)		890 (0.5)		485 (0.1)		980 (0.15)
	676 (0.05)	765 (0.1)		892 (0.5)		300 (0.1)		970 (0.05)
	695 (0.2)	775 (0.4)	815 (0.15)	875 (0.6)			842 (0.1)	982 (0.1)
		775 (0.2)	812 (0.3)	873 (0.5)			842 (0.5)	977 (0.25)
	704 (0.2)			885 (0.5)			842 (0.25)	982 (0.3)
				885 (0.7)			842 (0.1)	984 (0.2)
574 (0.3)				882 (0.2)			846 (0.6)	970 (0.1)
574 (0.4)				889 (0.5)	295 (0.2)		842 (0.2)	968 (0.1)

8. Zeolites

Among the framework silicates only some feldspars and the zeolites have been investigated differential thermal analytically. KÖHLER and WIEDEN describe the transformation from high- to low-temperature plagioclases between 780 and 820° C being a weak endothermic effect; in the author's DTA runs up to 1000° C, no thermal effects could be detected. *Zeolites* are alumosilicates built up to a bulky framework with hollow channels in the structure which can be filled up by water, and numerous easily interchangeable cations (Ca^{++}, Na^+ and others). The water can be released by heating without the structure being decomposed. Most of the zeolites lose their water below 400° C (see Table 25). Only laumontite, chabasite and the minerals of the natrolite group (natrolite, scolezite, thomsonite) dehydrate partly between 400 and 600° C, so demonstrating that a part of the water has been structurally bound more effectively than the water of the channels. The decomposition of

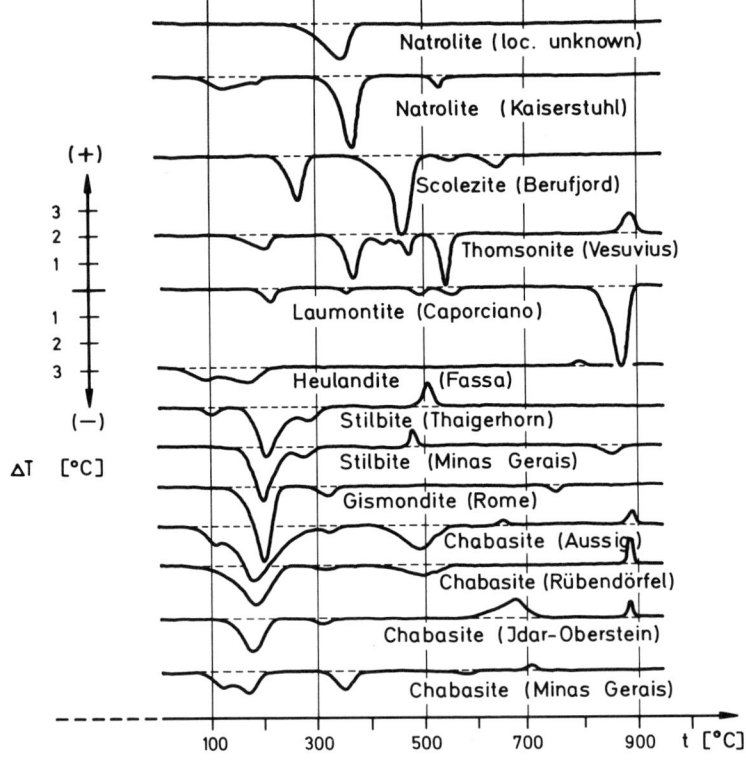

Fig. 50. DTA curves of zeolites

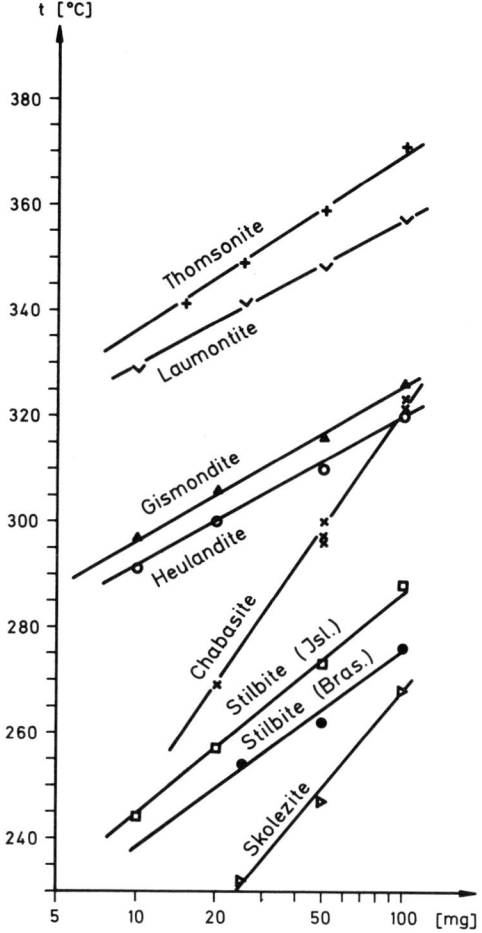

Fig. 51. PA-curves of some zeolites

zeolitic structures takes place between 520 and 750° C, with the exception of the minerals laumontite and stilbite, which are finally decomposed between 840 and 880° C. Many of the zeolites transform at higher temperatures (PÉCSINÉ-DONATH), e.g. natrolite in nepheline, chabasite in plagioclase. These transformations can be recognized by exothermic effects at 500° C (stilbite), 650–700° C (chabasite), 880–890° C (thomsonite and chabasite, see Table 25) and by a drifting of the zero-line into the exothermic range after 960° C. This reflects a process of sintering. The exothermic peaks demonstrate that the transformations are no phase

Table 25. DTA data of zeolites (in °C)

Mineral, origin, formula	Endothermic reactions (ΔT, °C)						Exothermic reactions (ΔT, °C)		
Natrolite, unknown $(Na_2[Al_2Si_3O_{10}] \cdot 2H_2O)$	125 (0.4)		350 (1.3)						
Natrolite, Kirchberg/Kaiserst., Germany, 50 mg!		190 (0.2)	370 (2.65)						
Scolezite, Berufjord/Iceld. $Ca[Al_2Si_3O_{10}] \cdot 3H_2O$		268 (1.7)			465 (3.0)	531 (0.4)	645 (0.45)		
Thomsonite, Vesuvius, Italy $NaCa_2(Al_2(Al,Si)Si_2O_{10}]_2 \cdot 6H_2O$		208 (0.5)	371 (1.65)	425 (0.15) 449 (0.1)	477 (0.8)	550 (0.15) 544 (2.0)		889 (0.75)	
Laumontite, Caporciano, Italy $Ca[AlSi_2O_6]_2 \cdot 4H_2O$	97 (0.4)	215 (0.5)		357 (0.15)	493 (0.2)	557 (0.2)	870 (3.05)		
Heulandite, Fassa/Tyrol $Ca[Al_2Si_7O_{18}] \cdot 6H_2O$		176 (0.5)			470 (0.5)			392 (0.2) 787 (0.2)	887 (0.1)
Stilbite, Theigerhorn/Iceld. $Ca[Al_2Si_7O_{18}] \cdot 7H_2O$	100 (0.2)	204 (1.9)	288 (0.5)					510 (0.95)	
Stilbite, Minas Geraes/Brasil		200 (2.1)	276 (0.35)				850 (0.35)	481 (0.6)	
Gismondite, Rome $Ca[Al_2Si_2O_8] \cdot 4H_2O$		201 (2.9)	326 (0.35)				745 (0.25)		
Chabasite, Aussig/CSSR $(Ca,Na_2)[Al_2Si_4O_{12}] \cdot 6H_2O$	108 (0.15)	176 (1.6)	321 (0.2)		495 (0.9)	520 (0.4)		650 (0.15)	882 (0.5)
Chabasite, Rübendörfel/Silesia		183 (1.5)	323 (0.1)		495 (0.35)	520 (0.25)			889 (1.0)
Chabasite, Idar-Oberstein, Germany		178 (1.25)	312 (0.1)					673 (0.7)	886 (0.6)
Chabasite, Minas Geraes/Brasil	120 (0.6)	172 (0.8)	350 (0.7)			581 (0.1)		700 (0.15)	

transformations from one crystal structure into another, but reconstructions of crystal structures, after the original structure having been decomposed. Dehydration behaviour and exothermic peaks give characteristic DTA and PA-curves of the minerals (Figs. 50 and 51) which allow a good determination of zeolites by means of DTA.

There are many contradictory results in the numerous publications dealing with the thermal behaviour of zeolites (see KOIZUMI; KOIZUMI and ROY; MASON and SAND; GRANGE; HARADA and TOMITA; MERKLE and SLAUGHTER; IJIMA and HARADA; KEUSEN). The reason for these contradictory results can be partly that the minerals have not been determined very exactly. For instance, the author's DTA investigations on natrolites corroborate the data of SVESHNIKOVR and KUZNETSOV (1946: main endothermic effect at 350° C) but not the recent data of PENG, who has found a main dehydration temperature of 455° C for natrolites. DTA data of chabasites are quite different in literature, too. The author's data, listed in Table 25, were obtained by standard conditions of analysis, the 13 samples have been identified by X-ray methods.

9. Allophane, Opal, and Organic Matter of Soils and Sediments

Allophanes are hydrated, X-ray amorphous alumosilicates of variable composition occurring in soils. Their thermal behaviour is similar to that of halloysites (WHITE), in which allophanes will be quickly transformed in nature by ageing and ordering (CHUKROV et al.). The endothermic dehydration peak between 130 and 185° C shifts a little towards higher temperatures with decreasing particle size (CAMPBELL et al.): the larger surface of finer particles is able to adsorb more water. This relation between peak temperature and surface of fine particles can be taken to estimate the grain size of fine-grained clay minerals (see III-5). Analogous to the DTA curves of halloysite and kaolinite, the curve of allophanes shows between 930 and 1000° C the exothermic peak reflecting the formation of spinel. The temperature of this exothermic peak decreases with increasing Fe-contents of allophanes (ΔT decreases, too) until the peak disappears at Fe/Al relations >0.3, in DTA curves of hisingerites, after RAMDOHR and STRUNZ allophanes containing iron. Allophanes are not found frequently in natural samples because they are quickly transformed to halloysite or fireclay. Therefore three samples of different chemical compositions have been prepared by simultaneous precipitation of $Al(OH)_3$ and SiO_2 (1) resp. of $(Al, Fe)(OH)_3$ and SiO_2 (2, 3)

Table 26. DTA data of allophane, opal and organic matter (°C)

Sample, origin	Chemical composition	Endothermic reaction (ΔT, °C)	Exothermic reaction (ΔT, °C)
"Allophane", synthetic	$SiO_2:Al_2O_3 = 1:2$	112 (1.6)	960 (0.6)
"Allophane", synthetic	$SiO_2:Al_2O_3:Fe_2O_3 = 1:1,5:0,2$	116 (1.8)	923 (0.2)
"Hisingerite", synthetic	$SiO_2:Al_2O_3:Fe_2O_3 = 1:1.3:0.4$	109 (1.5)	890 (0.05)
Opal, unknown	$SiO_2 \cdot n\,H_2O$	87 (0.3), 115 (1.4)	
Opal, unknown	$SiO_2 \cdot n\,H_2O$	62, 113, 142[a] (0.25) (1.4) (0.2)	
Limestone + earth wax ($\sim 1/10$ earth wax), Hils, Germany	$CaCO_3$ + organic matter	Not heated until $CaCO_3$ was decomposed	440 (3.5) 553 (3.8)
Limestone + traces of earth wax, Hils/Germany	$CaCO_3$ + traces of organic matter	Not heated until $CaCO_3$ was decomposed	472 (0.4)

[a] Inversion of strongly disordered cristobalite.

and drying at 50° C for one hour. The samples were then heated under standard conditions (Table 26).

Opals, $SiO_2 \cdot n\,H_2O$, naturally formed like allophanes by precipitation of gels and ageing, are X-ray amorphous minerals. They show in the DTA curve two or three endothermic peaks between 60 and 160° C that reflect the dehydration (the two first peaks being broad and very strong) and the sometimes appearing inversion peak of low- to high-cristobalite occurring in a few opaline substances. This third peak is weak but relatively sharp. Exactness of measurements $\pm 2°$ C; heating under standard conditions (Table 26).

The combustion of organic matter (coal, bitumen, earthwax etc.) which frequently occurs in soils and also in sediments, (e.g. in the posidonia slate of North Germany, see SCHMITZ and VON GÄRTNER), causes in the DTA curve an extremely strong exothermic deflection between 400 and 650° C. Very small amounts of organic matter can be proved from soil samples (see Table 26). SALGER has published DTA curves of whewellite, $CaC_2O_4 \cdot H_2O$, and other organic substances occurring in soils.

10. Development of Identification Diagrams

In another important investigation technique applied in mineralogical sciences, the optics, a first introductory diagram exists which shows refractive index versus double refraction. With the aid of this diagram a mineral can be searched for, of which the optical characteristics are known by measurements. In differential thermal analysis it will be much more difficult than in optics to develop a similar introductory diagram. But on principle it is possible, if we take into consideration that DTA data are strongly influenced by a lot of apparative and preparative factors. If we do so and if we give some additional remarks, we will be able to construct an introductory diagram for DTA, too, using the *temperature and ΔT as abscissa resp. ordinate* to figure the characteristic data of a DTA peak. But we must not forget that both characteristics, the peak temperature as well as ΔT, can only be comparable with the data of other DTA runs if all runs have been made under comparable heating conditions. The following consequence is: such an introductory diagram temperature versus ΔT of a peak can be only valid for one set of investigation conditions. If you vary the heating rate, for example, you will get a shifting of the data. Therefore I propose to work under the standard conditions of analysis as listed in Table 1, and for these conditions the diagrams figured in the present study will be valid.

Figuring the peak temperature on the abscissa, a first difficulty will be that we have to construct *not only one diagram but several introductory diagrams*, in order to get a general view of all thermal effects of the minerals being investigated. On looking at the following Figs. 52–60, you will see that it has been necessary to subdivide the temperature range from 20 to 1000° C into *nine diagrams*, seven for endothermic and two for exothermic peaks. But this will only be a difficulty in drawing.

The *parameter ΔT* has to be drawn on a *logarithmic scale*, since there are a lot of ΔT-values in the order of 0.1° C as well as in the order of 10° C. The measurement of ΔT will be done as follows: In the case of thermocouples of Pt-Pt$_{90}$/Rh$_{10}$ the temperatures 20, 100, 200, 300 ... 1100° C are represented by 0.0 mV; 0.53 mV; 1.33 mV; 2.21 mV; 3.15 mV; 4.11 mV; 5.12 mV; 6.16 mV; 7.23 mV; 8.34 mV; 9.49 mV; 10.66 mV. The thermal power of thermocouples are dependent on the temperature, but between 300 and 800° C (resp. 2.21 mV and 7.23 mV) this dependence is very slight, so demonstrating that 1 mV is adequate to nearly 100° C or 0.01 mV is adequate to 1° C. Now the size of a peak deflection has to be related to the scale of the DTA apparatus marked "sensitiveness of measurement", which generally amounts to values between 25 µV and 2 mV. In the case of heating with a sensitiveness of, say, 100 µV (=0.1 mV), a peak deflection over the whole range (from one edge of the recorder to the other) will be adequate to a ΔT-value of nearly 10° C, and a small peak only going over a hundredth of the recorder scale will then be adequate to a ΔT-value of 0.1° C. Generally this will be the lower margin of ΔT-measurement. In this way the parameter ΔT will be obtained by measuring from the DTA zero line. For very exact ΔT-values you have to consider the temperature dependence of the thermal power of the thermo-couples by correcting the measured values by a calculated factor of this temperature dependence. Generally, however, this correction will not be necessary, because these diagrams of peak temperature versus ΔT will only be used as "introductory", as *key diagrams* for the *identification of an unknown mineral*.

The values peak temperature versus ΔT obtained by investigating an unknown mineral in the DTA under standard conditions, however, will correspond to points in the figured diagrams (Fig. 52–60), or to points on the lines in these diagrams, or they will lie at least very close to such a line (or point), provided that the unknown mineral has been investigated once before. That means: its values peak temperature versus ΔT will be contained in the figured diagrams. Let me give an example. A DTA curve shows two endothermic effects at 120 and 563° C and one exothermic peak at 975° C. According to Fig. 52 the peak at 120° C can be related to the dehydration of chabasite, palygorskite, vermiculite, sepiolite, halloysite, montmorillonite, nontronite or soda and to the melting of sulfur. The ΔT-value of 1.5° C indicates the presence of halloysite, and the data of the two other peaks at 563 and 975° C corroborate the presence of this mineral.

On the diagrams (Figs. 52–60), a lot of lines generally beginning on the bottom left and ending on the top right, going from lower temperature and ΔT-values to higher ones are visible. This can be explained by crystal physical and crystal chemical differences of the minerals (compare with part III of the present study): the peak temperatures as well as the ΔT-values are dependent on the chemical composition and on the degree of disorder of the minerals; this effect is very remarkable in clay minerals and sedimentary micas (illites, glauconites), while the data of carbonates, hydroxides, borates, sulfates and phosphates vary only little from the given points in the figures. The exothermic peaks vary only in their ΔT-values (see Fig. 60). If the DTA curve of an unknown mineral shows more peaks, the idenfication of this mineral will be much easier (see above: halloysite). Very suitable for identification of minerals by DTA are exothermic peaks (as in sulfides and sheet silicates), and endothermic peaks reflecting structural transformations occurring very spontaneously (like in some sulfides, oxides and carbonates).

The fact of the sample being a *mixture* of some minerals can be recognized either by the occurrence of the DTA effects of several minerals in the DTA curve, or by big deviations of the temperature and ΔT-values from the given values of diagrams 52–60. In this case the obtained data and curves must be compared with all DTA data and curves of the minerals which lie close to the values of the unknown sample in the Figs. 52–60. Mostly this comparison will lead to the identification of the unknown mineral. In some cases one may have to consider additional DTA information for this identification, e.g. the a-values in the case of clay minerals (see II-7.8). For samples consisting of mixtures of several minerals, however, the Figs. 52–60 will only serve as key diagrams.

Naturally the present study only gives a selection of all existing minerals which can be identified or even investigated by DTA. But among the 148 minerals listed below and investigated in this study, you will find all the minerals which frequently occur in sediments, soils, igneous and metamorphic rocks, provided that they show any thermal effect. Using the proposed standard conditions (Table 1), it will be possible at any place and at any time to complete the diagrams by your own analysis, added to the given data.

A comparison of our DTA data with those of HURLBUT and ARISTARAIN (rare borate minerals) or of HEIDE (sulfates) shows that in cases of comparable conditions of analysis, such comparison can at least lead to the *group* of minerals, to which an unknown belongs, and within this group to a mineral that is similar in its thermal behaviour and thus reflects crystallographical and chemical similarities, too. So, for example, some of the rare borates investigated by ARISTARAIN and HURLBUT can well be compared with borates in the present study: the DTA data of rivadavite, ezcurrite and other borates taken from curves published by ARISTARAIN and HURLBUT resp. HURLBUT and ARISTARAIN would lie very close to the points of kernite and colemanite in the Figs. 52–60. It would be analogous to compare our data of sulfates with those of HEIDE, the data of carbonates with those of BECK, the data of clay minerals and hydroxides with those of MACKENZIE, PETERS and others. The following diagrams 52–60 contain the data of 148 minerals marked by different symbols: + elements and sulfides; □ halogenides and sulfates; △ oxides and hydroxides; ▽ borates, phosphates and arsenates; ○ carbonates and nitrates; × sheet silicates; ⊗ other silicates.

The abbreviations stem from the German mineral names. In most cases German and English names are similar, the exceptions can easily be found in the following list (e.g. "Glimmer" = mica), which has been ordered according to the alphabetical sequence of the abbreviations.

Table of Abbreviations Used in Figs. 52–60

Ak	acanthite	Fire	fireclay-mineral
Al-Ch	Al-chlorite	G	goethite
Al-G	Al-goethite	Gay	gaylussite
Alp	allophane	Gb	gibbsite
Alu	alunite	Ger	germanite
Am	amphibole	Gi	gypsum
Amb	amblygonite	Gis	gismondite
An	annabergite	Gla	glauconite
Ank	ankerite	Gli	mica
Ar	aragonite	Gos	goslarite
Art	artinite	Hä	hematite
As	astrakanite	Hai	haidingerite
Au	aurichalcite	Hal	halloysite
Az	azurite	Hau	hausmannite
Bas	bastnaesite	Hec	hectorite
Bay	bayerite	Hem	hemimorphite
Bio	biotite	Heu	heulandite
Bö	boehmite	Hun	huntite
Bor	bornite	Hym	hydromagnesite
Borax	borax, tincal	Hyt	hydrotalcite
Bou	boulangerite	hyz	hydrozincite
Bre	breunnerite	Il	illite
Brg	brugnatellite	Kao	kaolinite
Bru	brucite	KCl	sylvite
C	graphite	Ker	kernite
Cc	calcite	KNO_3	nitre
Ch	chlorite	Kry	cryolite
Chal	chalcopyrite	Lau	laumontite
Chr	chrysocolla	Le	lepidocrocite
Chry	chrysotile	Li	limonite
Col	colemanite	Ma	malachite
Cor	corrensite	Mag	magnetite
Cr	cristobalite	Mar	marcasite
Cu	chalcocite	Mel	melanterite
Di	dickite	Mg	magnesite
Dia	diaspore	Mg-Ch	Mg-chlorite
Dol	dolomite	(Mg,Fe)-	
Ery	erythrite	Ch	Mg-Fe-chlorite
F	fluorite	Mi	mirabilite
Fe-Ch	Fe-chlorite	Mn	manganite

Mn-C	Mn-calcite	Sep	sepiolite
Montm	montmorillonite	Si	siderite
MoS$_2$	molybdenite	Sk	schröckingerite
Ms	muscovite	sko	scolezite
NaCl	halite	Sm	smithsonite
Nah	nahcolite	Soda	soda
Nat	natrolite	SrCO$_3$	strontianite
Nes	nesquehonite	St	stephanite
Non	nontronite	Sti	stilbite
Nor	norsethite	Str	struvite
Op	opal	Ta	talc
Pal	palygorskite, attapulgite	Ten	tennantite
		Th	thomsonite
Par	parisite	The	thenardite
Pb-Cc	plumbocalcite	Tr	tridymite
PbS	galena	Tro	trona
Pho	phosgenite	Turm	tourmaline
Pir	pirssonite	Ul	ulexite
Pol	polyhalite	Verm	vermiculite
Py	pyrite	Verm(g)	well ordered vermiculite
Pyl	pyrolusite		
Pyo	pyrrhotite	Vi	vivianite
Pyp	pyrophyllite	Vo	voglite
Q	quartz	Wi	witherite
Q(f)	quartz, disordered	Wu	wurtzite
S	sulfur	Za	zaratite
Sap	saponite	Zb	sphalerite
Sas	sassolite	Zi	zinnwaldite
Ser	serpentine		

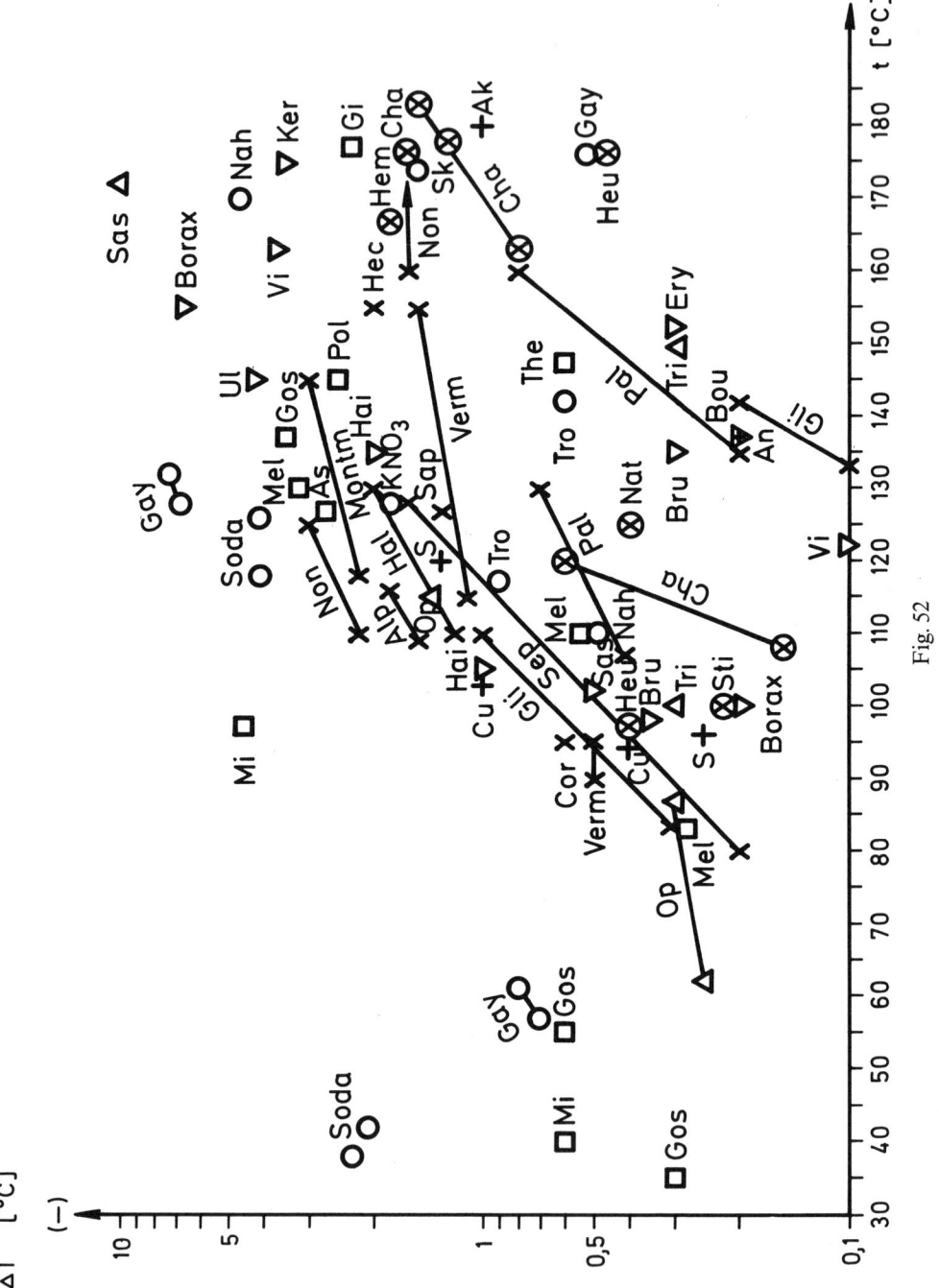

Fig. 52

Identification Diagrams

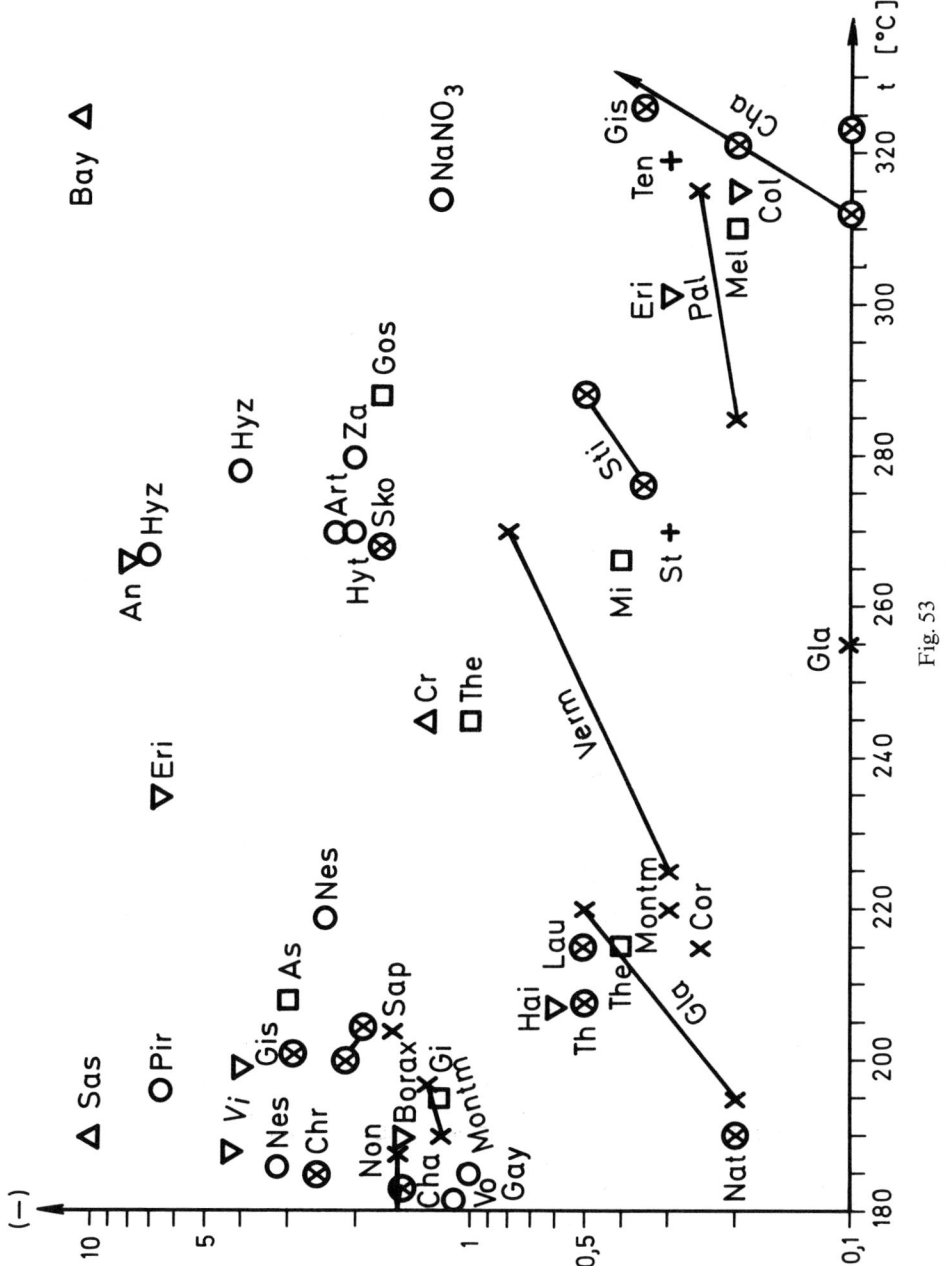

Fig. 53

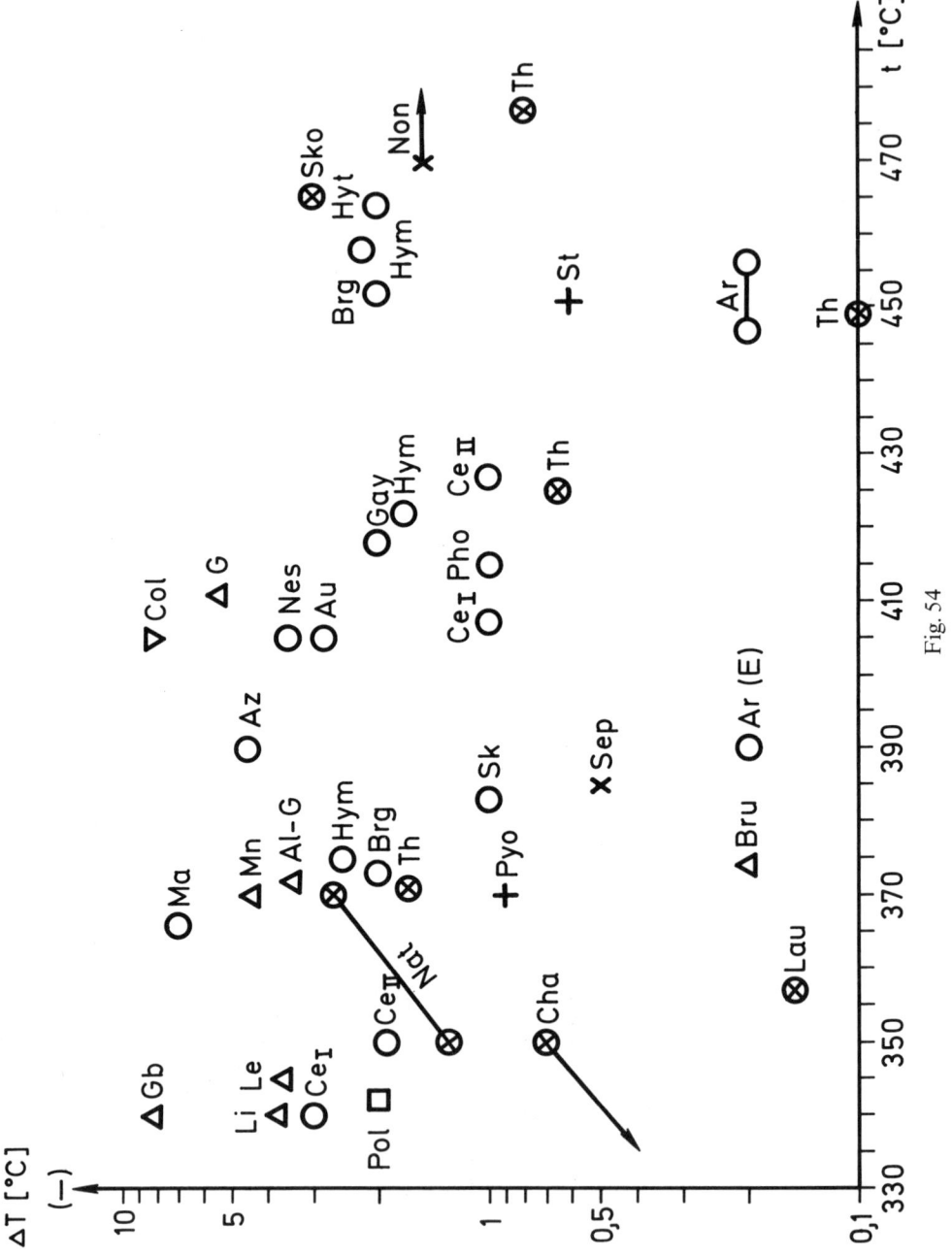

Fig. 54

Identification Diagrams

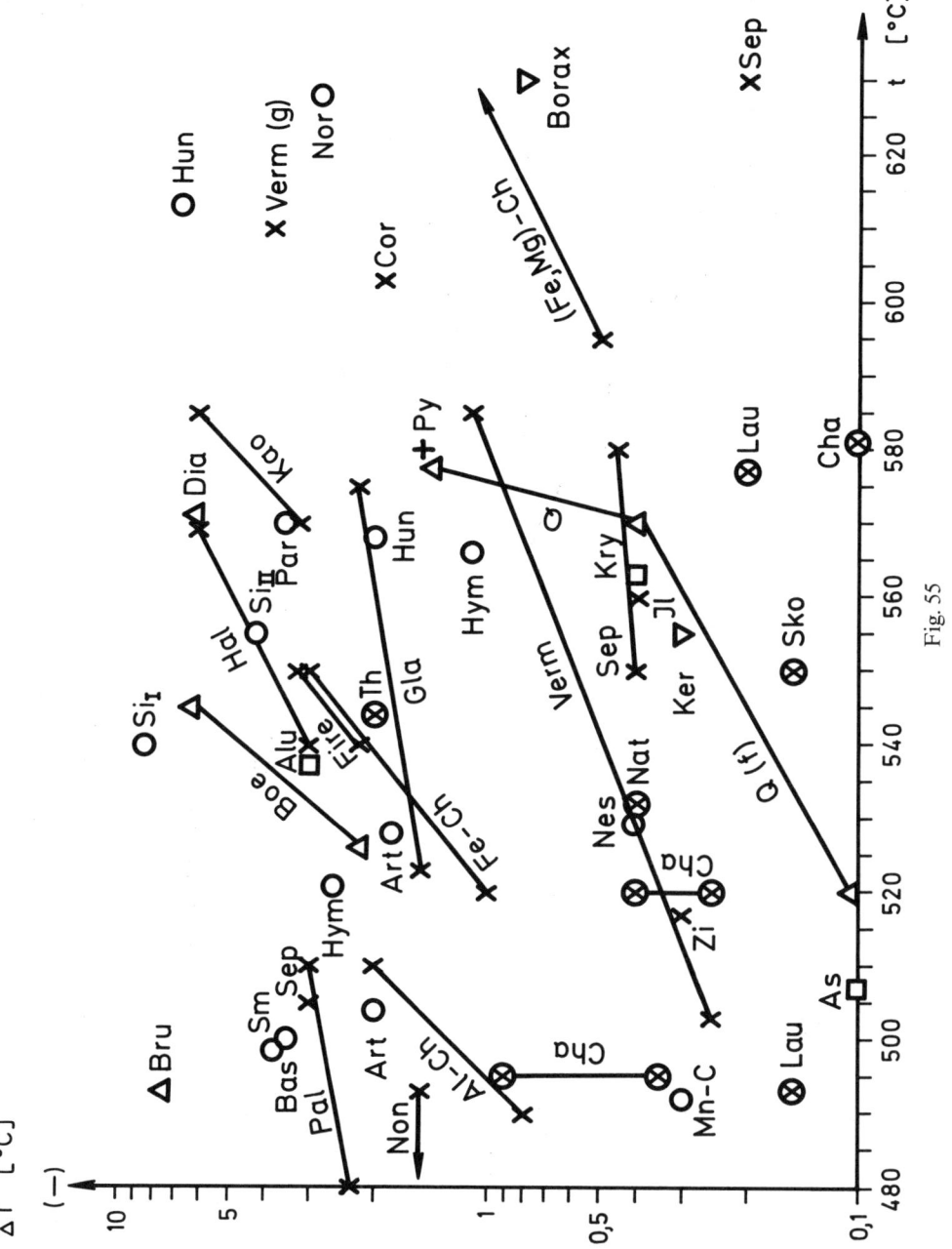

Fig. 55

102 Part II. Application of Differential Thermal Analysis

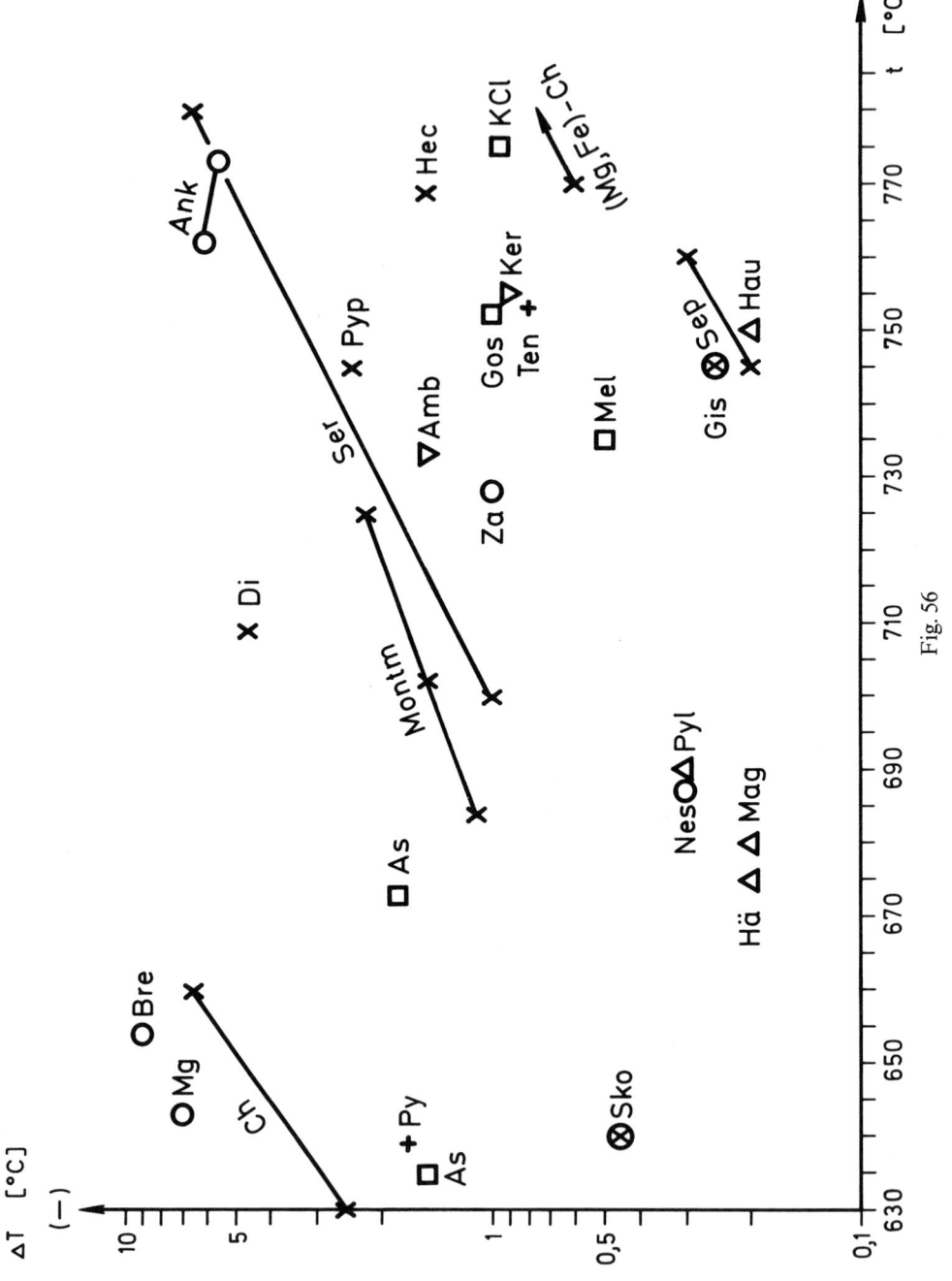

Fig. 56

Identification Diagrams

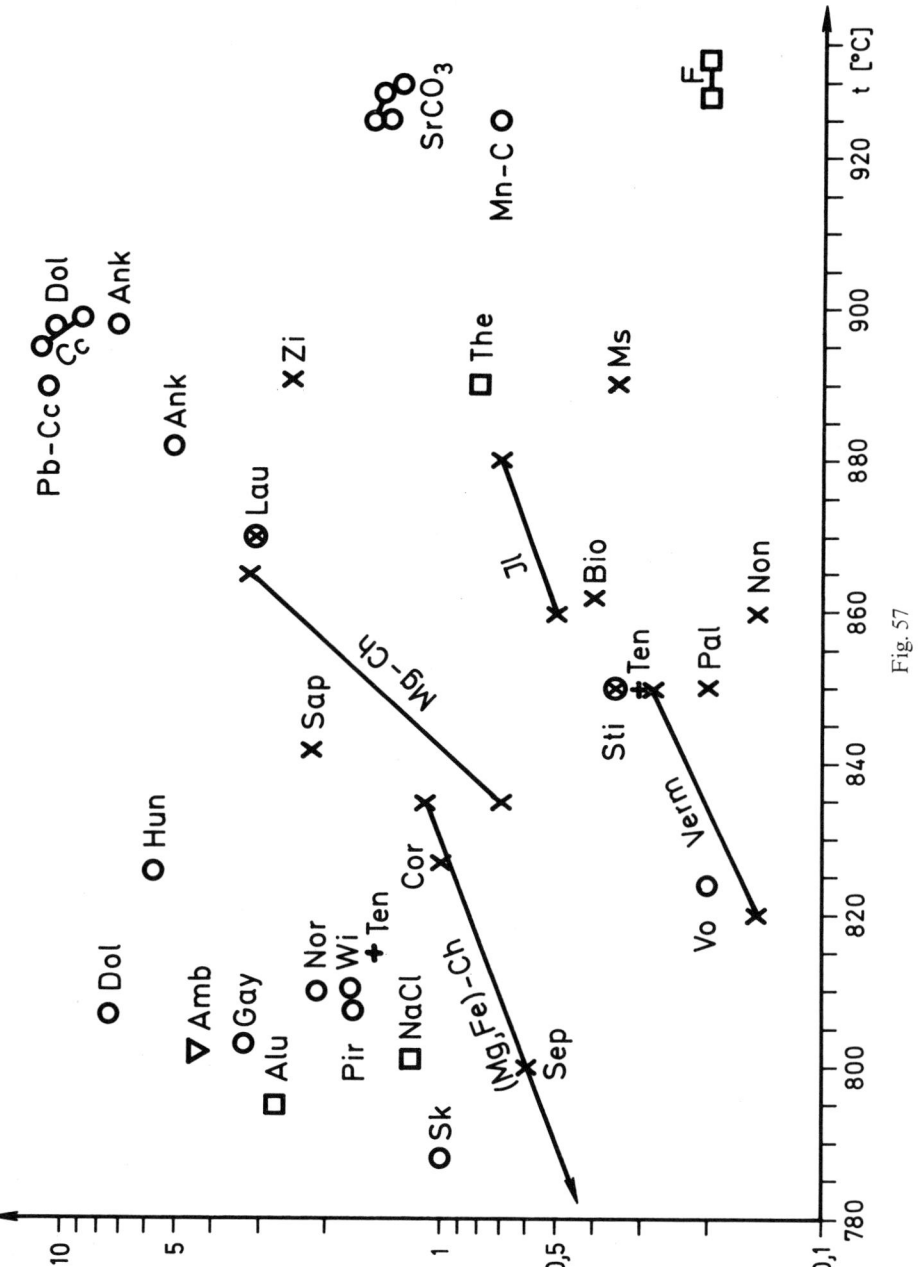

Fig. 57

104 Part II. Application of Differential Thermal Analysis

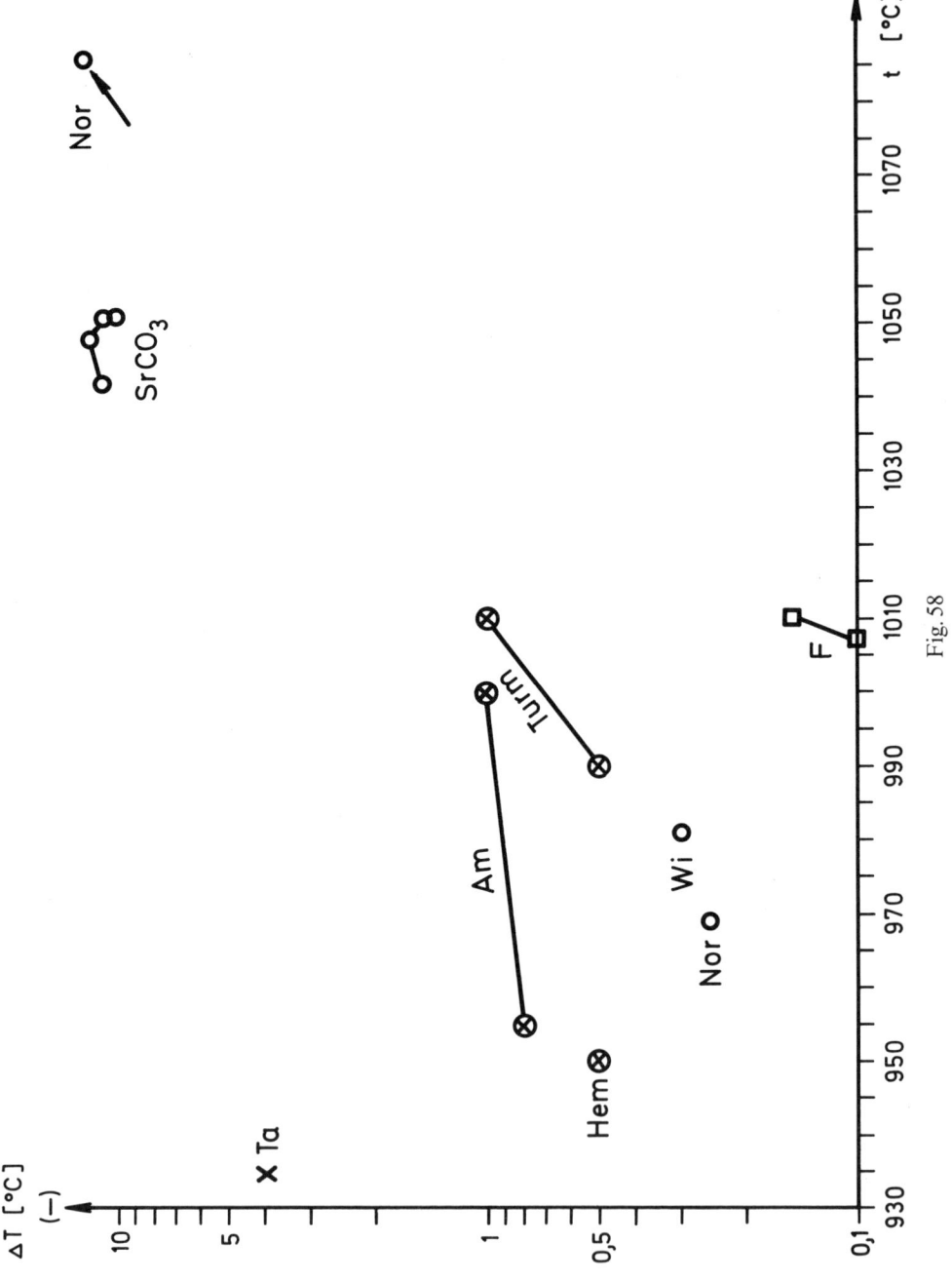

Fig. 58

Identification Diagrams

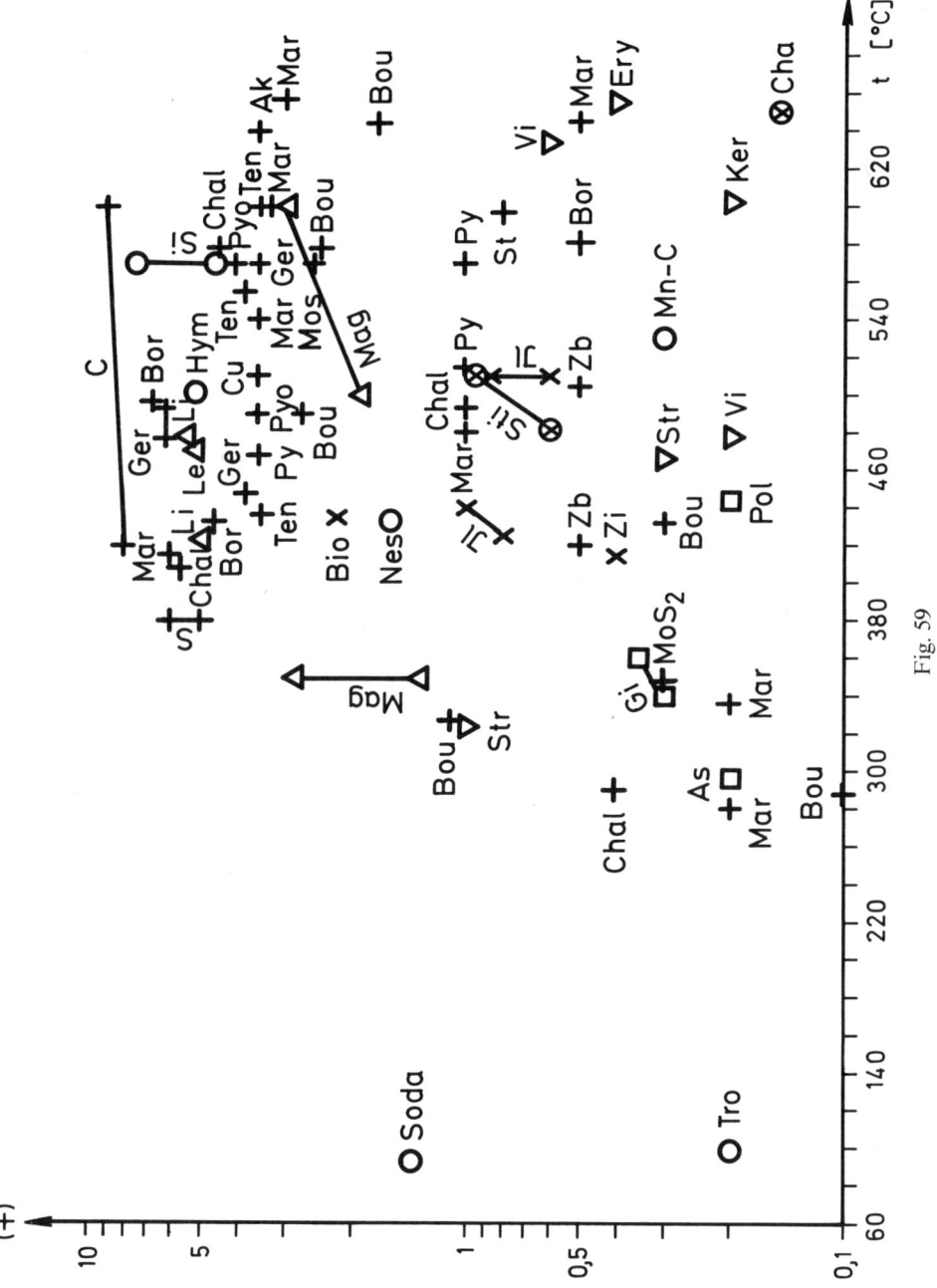

Fig. 59

106   Part II. Application of Differential Thermal Analysis

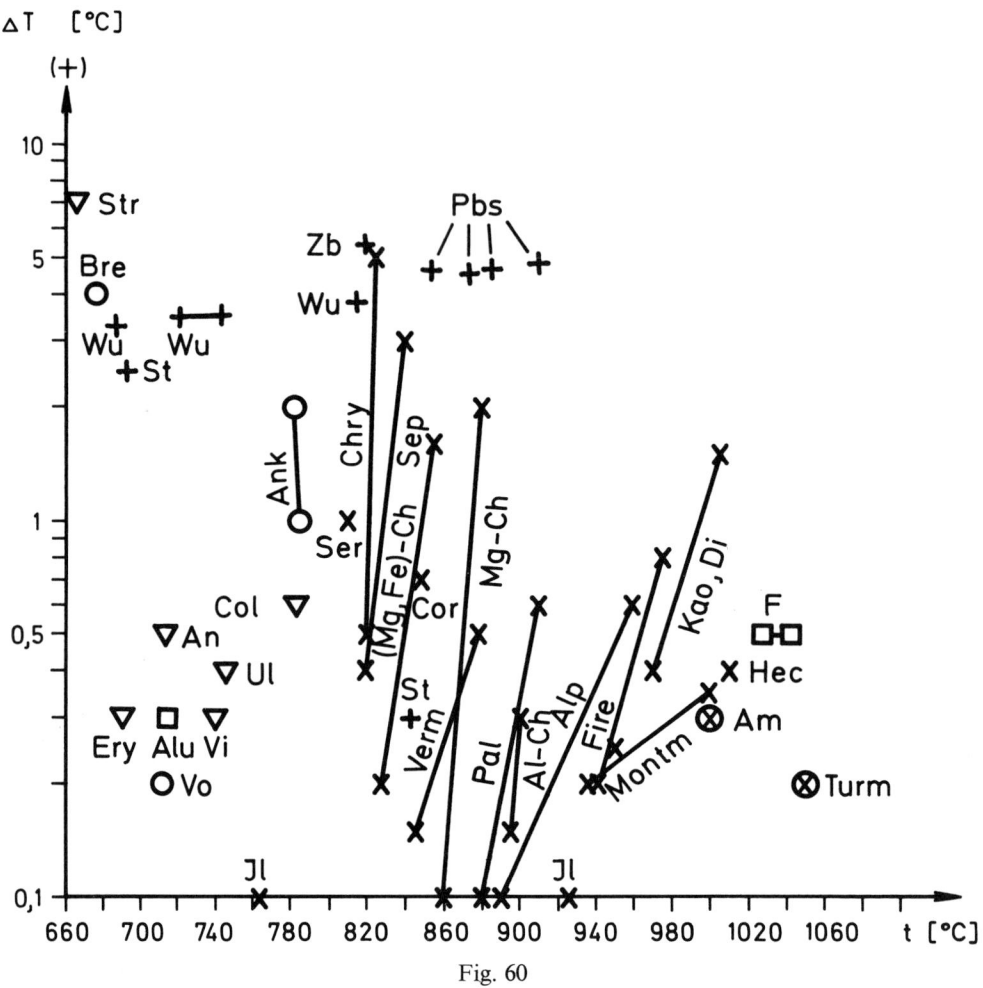

Fig. 60

Part III. Special Application of Differential Thermal Analysis in Mineralogy: Statements about Chemical Composition, Degree of Disorder, and Genesis of Minerals

1. Influence of the Chemical Composition on the Decomposition Temperatures of Carbonates and Hydroxides

The *decomposition* of a pure, scarcely disordered crystal by means of pressure or heat will occur very spontaneously: the crystal structure will keep stable until a sufficient amount of energy has been supplied, then breaking down at many spots simultaneously. This will be reflected in the DTA curve by a very sharp endothermic peak with relatively large temperature difference (ΔT). However, if the crystal contains some impurities or crystal physical defects, at these spots of disorder the stability of the structure is decreased. These disordered parts of the crystal need a lower supply of energy being destroyed than other parts of the crystal which are not disordered. That means: the decomposition of the crystal will begin earlier, starting at these spots of disorder being reflected in the DTA curve at a lower initial temperature and a lower value of ΔT. Compared with the sharp decomposition peak of a pure crystal, the DTA curve of the same mineral that is impure or disordered only shows a broad deflection of lower intensity without a distinct endothermic minimum. The *shape (form)* of a decomposition peak therefore indicates if substitutions or crystal physical defects are present or absent in a crystal structure. This statement will be possible if there is no coincidence of two or more thermal effects, as in the case of simultaneous dehydration and decomposition. In DTA curves of hydroxides or clay minerals the dehydration and decomposition effect generally overlap. This can be seen in a broad, well-rounded endothermic deflection, bringing into disorder the relation between decomposition temperature and stability of the structure. This relation can be better seen in DTA curves of minerals free of water. The distinction between a decomposition peak influenced only by chemical impurities (e.g. substitutions of ions by others) and another only (or mainly) influenced by crystal physical defects is very difficult. But it will be possible to distinguish between these two kinds of effects influencing the stability of the crystal structure if you are able to make some additional suggestions either by means of X-ray and optical investigations or by DTA itself. So the estimation of the influence of chemical impurities (substitutions) on the decomposition peak temperature can be made if a crystal can be regarded as well ordered by means of other investigation methods. Under some fortunate circumstances, strange elements (ions) incorporated in a crystal structure can be recognized in the DTA curve by a separate endothermic peak. This will indeed be very small but very clear, too, as is the case in curves of aragonites containing some Sr^{++} or Ba^{++} at the Ca^{++} position of the structure.

1.1 Substitution of Ca^{++} by Mg^{++} or Pb^{++} in Calcites

Calcites occurring in hydrothermal Pb-Zn veins can incorporate Pb up to 0.4 mole-% (compare with the "plumbocalcite from Bleiberg", II-4). Recent carbonate sediments contain Mg-calcites which show a substitution of Ca^{++} by Mg^{++} up to 25 mole-% Mg. Since the ionic radii of Mg^{++} and Ca^{++} are different (0.78 resp. 1.06 Å), the substitution of Ca^{++} by Mg^{++} lowers the stability of the calcite structure. The first recrystallization which takes place during early diagenesis will stabilize the calcite structure by removing the Mg^{++}, being no longer tolerated in the structure. Therefore Mg-calcite can only be found in recent carbonate sediments and in relics of young organisms (see Table 27). Actually Pb^{++} is too large to be substituent in the calcite structure (ionic radius of Pb^{++}: 1.32 Å!); the low Pb^{++}-content being available in few calcites, it lowers the stability of the calcite structure remarkably (see (Fig. 18).

Table 27. MgO contents and decomposition temperature of Mg-calcites[a] (150 mg)

Sample, locality	Mole-% MgO	Decomposition temp. (°C ± 0.5°)	$d_{(10\bar{1}4)}$, Å
Pure calcite, Harz/Germany	0	925	3.035
Mg-calcite, Bermudas	5.9	898	3.018
Encrinus liliformis, Bermudas	8.65	893	3.005
Calcitic algae, Gaybu	18.1	889	2.974
Calcitic algae, Sao Vincente	19.0	889	2.972
Calcitic algae, Fernando Poo	21.0	887	2.966

[a] The samples were kindly supplied by Prof. C. W. CORRENS, Göttingen.

The X-ray interferences of five Mg-calcites studied were very sharp, as the DTA peaks also were, so characterizing the samples as well-ordered calcites. The interferences had, of course, shifted a little to lower d-values. The data of Table 27 are valid for an amount of the sample material of *150 mg* heated in a nickel block sample holder, otherwise under standard conditions.

The resulting curve in a diagram decomposition temperature versus mole-% Mg incorporated in the calcite structure can be used very well as a determination curve for Mg^{++} in calcites at Mg-contents <8 mole-%. At higher Mg-contents than 8 mole-% the peak temperature will only be very slightly lowered. (Compare with SMYKATZ-KLOSS, 1966.)

1.2 Substitution of Ca^{++} by Sr^{++}, Ba^{++}, and Pb^{++} in Aragonites

The structure of aragonite, the orthorhombic modification of $CaCO_3$, tolerates large divalent ions other than Ca^{++} occupying the Ca^{++} positions of the structure to an amount of some percent, e.g. Sr^{++} (ionic radius: 1.27 Å), Ba^{++} (1.43 Å) or Pb^{++} (1.32 Å). When the contents of these ions exceed 0.2 weight-%, in the DTA curve a small endothermic shoulder *before* the $CaCO_3$ decomposition peak and two sharp little peaks *after* this decomposition peak will appear due to the decomposition of the Pb-, Sr-, and Ba-carbonate components of aragonite. FAUST, who made the same observation (FAUST, 1950), explained these little peaks after the $CaCO_3$ decomposition as showing the retarded dissociation of some coarser grains of calcite. The author has discussed elsewhere why that cannot be (SMYKATZ-KLOSS, 1964). These little peaks do indeed reflect the decomposition of the (Sr-, Ba-, Pb-) carbonate components of aragonite, as has been corroborated by WEBB (personal information), who investigated aragonites with known Sr^{++} contents. Since in X-ray diagrams of aragonites containing 0.5 and 0.7 weight-% Sr^{++} (according to chemical analysis) no admixtures of strontianite could be detected, these Sr^{++} should be incorporated in the aragonite structure. Possibly in an aragonite crystal there may exist parts rich in Sr (Ba) and with a structural stability higher than that of the host crystal which are decomposed after the destruction of the $CaCO_3$ structure. In Fig. 61 two little peaks due to Pb^{++} and Sr^{++} are marked by arrows. The small endothermic peak around 450° C in the DTA curve of aragonites is due to the conversion aragonite→calcite (see II-4.1). The diagram for determining Sr^{++} in aragonites (Fig. 62) is valid for 150 mg heated in a nickel sample holder under standard conditions. The data of Fig. 62 should not be influenced by additional Ba^{++} or Pb^{++} contents (for detailed discussion see SMYKATZ-KLOSS, 1964).

110   Part III. Special Application of Differential Thermal Analysis

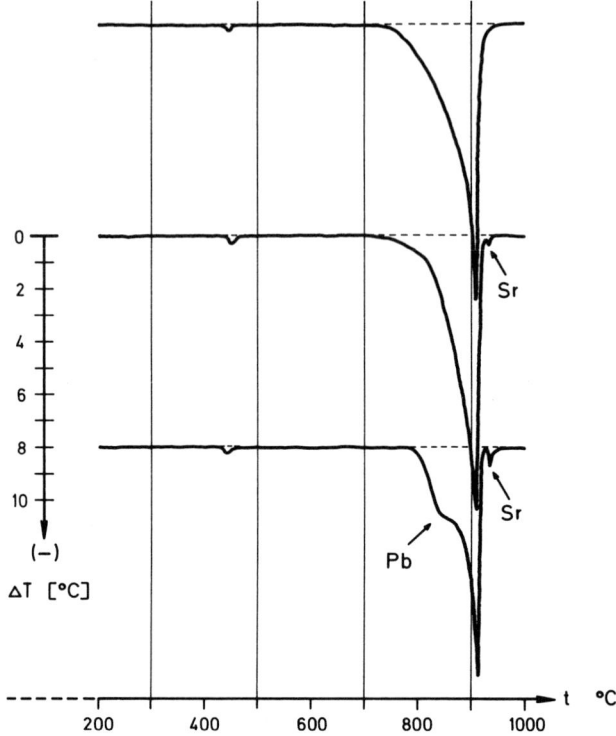

Fig. 61. DTA curves of some aragonites

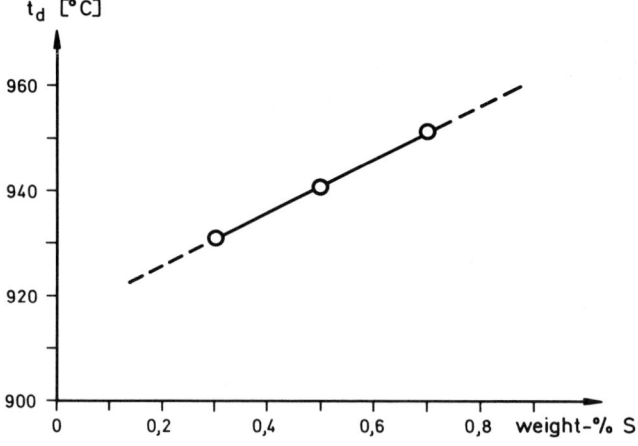

Fig. 62. Diagram for determining Sr^{++} in aragonites

1.3 Substitution of Mg^{++} by Fe^{++} and Mn^{++} in Dolomites

The decomposition temperatures of natural dolomites from recent carbonate muds vary about 30° C due to traces of soluble salts (mainly NaCl) having precipitated on the surface and in clefts of dolomite grains (GRAF). But older dolomites no longer in contact with sea water are free of salts. Therefore differences in the shape or the temperatures of the decomposition peaks of standardized DTA curves can only be related to the variation of the chemical composition of dolomites, provided that there are no perceptible crystal physical defects which can also be responsible for a variation of the decomposition deflections. The comparison of chemical analysis, X-ray investigations and other methods on the one hand with the thermal behaviour of some hundred dolomites on the other hand, leads to the conclusion that the substitution of Mg^{++} by Fe^{++} and/or Mn^{++} is responsible for such a variation of shape and temperature of endothermic peaks. In DTA curves of iron bearing dolomites a third endothermic effect before the first decomposition peak of dolomites can be observed, forming a shoulder hardly visible in the curves of low-iron dolomites or a separate peak in the curves of dolomites rich in iron (see Fig. 63: the "iron peaks" have been marked by arrows). If besides a remarkable Fe-content *manganese* is also present as substituent, even a fourth endothermic effect can be detected. This lies just between the first endothermic peak due to the decomposition of the $FeCO_3$-component of the dolomite, and the deflection around 800° C due to the decomposition of the $MgCO_3$-component. So the dolomite containing Fe^{++} and Mn^{++}, $Ca(Mg, Fe^{++}, Mn^{++})(CO_3)_2$, is decomposed in four steps. The dolomite structure is stable until the $CaCO_3$-component has been decomposed, too (the 4th step). That means: a dolomite heated up to 850° C has lost nearly half of its original CO_2 but is still a dolomite showing the X-ray interferences characteristic for this mineral. The different decomposition temperatures of these four components of the dolomite structure ($CaCO_3$, $MgCO_3$, $FeCO_3$, $MnCO_3$) can only be explained by differences in the bonding strength between the cations and CO_3^{2-}. In the case of dolomites containing remarkable amounts of Fe^{2+} and Mn^{2+}, both peaks due to the decomposition of the Fe- and Mn-component disturb each other (K. L. REDDICK, personal communication), but dolomites free of Mn^{++}, which incorporate only the substituent Fe^{2+}, can be investigated by DTA in order to determine the Fe-content (see Fig. 64a and b and compare with SMYKATZ-KLOSS, 1966).

This separate decomposition peak of the Fe-component increases in its peak area but decreases in peak temperature with increasing Fe-content (see Fig. 63), so shifting towards the decomposition peak of pure

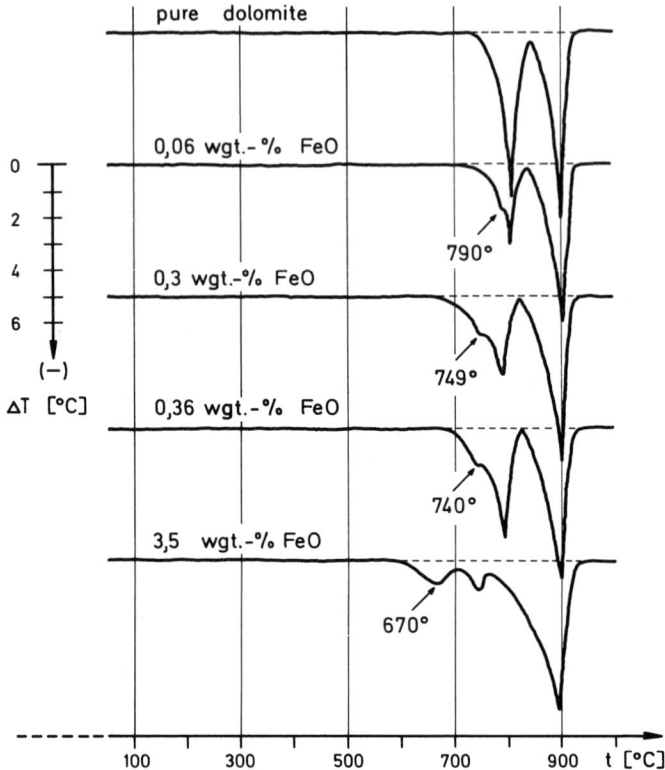

Fig. 63. DTA curves of some Fe-bearing dolomites

siderite, $FeCO_3$, which lies around 570° C (150 mg!). In the case of absence of Mn^{++}, the diagram peak temperature ($FeCO_3$-component) versus weight-% FeO incorporated in the dolomite structure can serve as a diagram for determining Fe-contents lying between 0.1 and 0.5 weight-% FeO (see Fig. 64b). Fe-contents higher than 0.5 weight-% FeO can only be determined inaccurately by DTA (see Fig. 64a). Many natural dolomites, however, above all those having been formed during early diagenesis like the dolomites from the Zechstein or Trias of South- and North-Germany, are low-iron dolomites (FeO <0.5 weight-%, see SMY-KATZ-KLOSS, 1966), suitable for Fe-content determinations by means of DTA. Heating occurs under standard conditions, with the exception of the amount of sample material (150 mg) and the use of a nickel sample holder.

If the incorporation of Fe^{++} and Mn^{++} into the dolomite increases to large amounts, forming the minerals ankerite, $CaFe(CO_3)_2$, or kut-

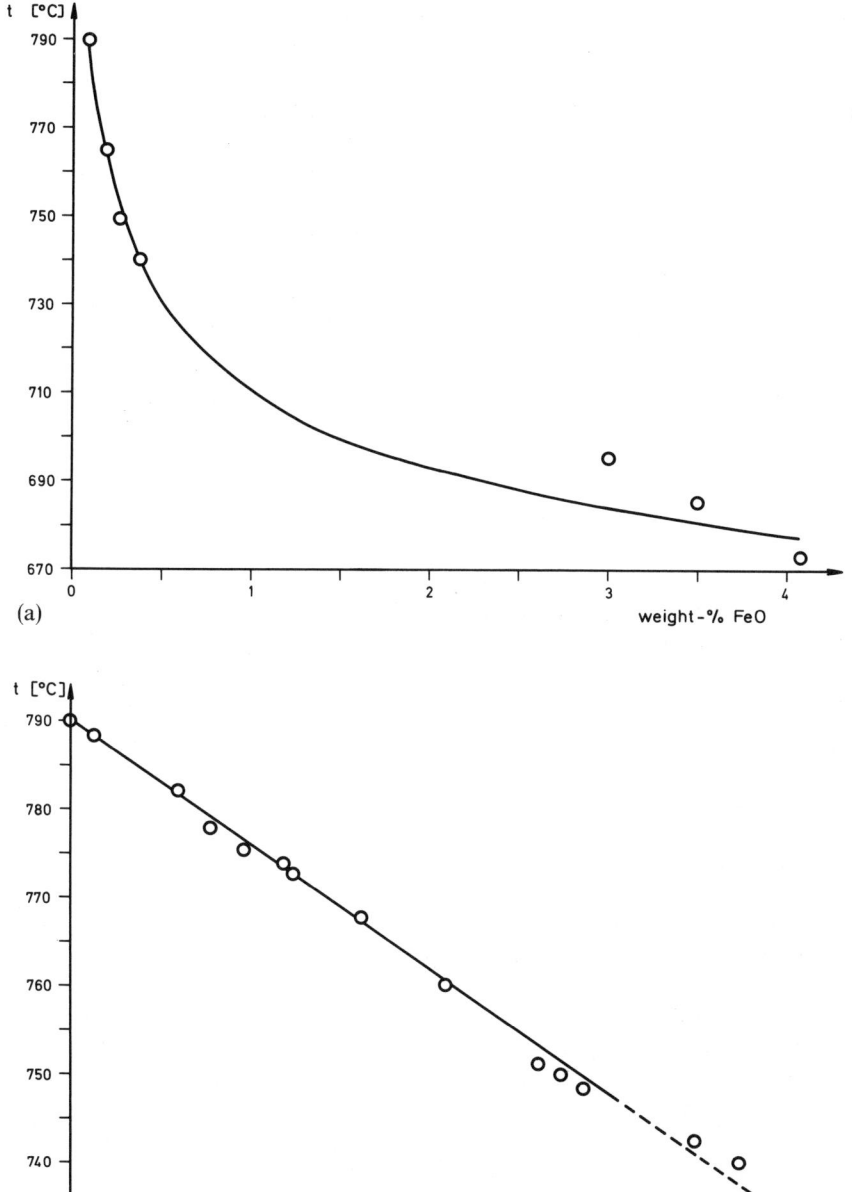

Fig. 64a and b. Diagram for determining Fe^{++} incorporated in dolomites

nahorite, $CaMn(CO_3)_2$, the endothermic decomposition peaks will be badly disturbed by an overlapping with the exothermic oxidation effects ($Fe^{2+} \to Fe^{3+}$; $Mn^{2+} \to Mn^{4+}$). In the case of ankerites and kutnahorites DTA can only be used for coarse qualitative investigations, but not for (semi-) quantitative determinations as described in II-4.1 or in this chapter.

1.4 Hydrozincite and Aurichalcite

Hydrozincite, $Zn_5[(OH)_3CO_3]_2$, can incorporate Cu^{++} into its structure, so forming aurichalcites, $(Zn,Cu)_5[(OH)_3CO_3]_2$, which have a somewhat higher structure stability than hydrozincite. The aurichalcite structure being orthorhombic disphenoidal can be explained by a simple twinning of the hydrozincite lattice, which is monoclinic prismatical. The decomposition temperatures of both minerals hydrozincite and Cu-rich aurichalcite differ by about 130° C; an aurichalcite lacking in Cu has been decomposed at 290° C (100 mg), nearly 20–25° C higher than pure hydrozincite. Probably the diagram decomposition temperature versus copper content will give a suitable determination curve for Cu in aurichalcites, but no useful samples have been available to carry on this problem. The same applies in the case of other carbonates forming solid solutions: for numerous carbonate minerals DTA should be a method very reliable for studying the dependence of structural stabilities on the chemical compositions, e.g. on various substitution phenomena as described in this chapter.

1.5 The Incorporation of Al^{+++} into the Structure of Goethite

The combined dehydration and decomposition peak of goethite, FeOOH, is lowered in its temperature when increasing Al-contents are incorporated into the goethite structure, where they substitute the trivalent iron. But the measurement of such lowering of peak temperatures of hydroxides caused by substitutions is much more difficult than for carbonates, and will only be successful in combined differential thermal, grain size and X-ray investigations. These will ensure that samples are comparable with those used for calibrating. The hydroxide minerals from various localities frequently differ in grain size and also in degree of disorder. Moreover, the lowering of the decomposition/dehydration peak temperature with increasing substitution generally amounts to no more than 20 or 30° C. This is the conclusion of the investigation of a series of synthetic goethites (samples from THIEL) incorporating different contents of Al. Therefore differences in peak temperatures of natural goethites (or other natural hydroxides) can be explained not only by different degrees of disorder or grain size differences, but also by varying amounts of incorporated substituents like Al in the iron hydroxide mineral goethite.

2. Influence of the Chemical Composition on the Temperatures of Structural Transformations

2.1 Carbonates

While differences in the temperatures of the monotropic aragonite→-calcite transformation (387–460° C, see Table 10) in no way show any relation to variations of the chemical composition, e.g. to the Ba-, Sr- or Pb-contents of these aragonites, in the case of *reversible* structural transformations (inversions), this relation can be observed. So witherite, $BaCO_3$, shows two minor but clearly observable endothermic effects at $811 \pm 2°$ C and $981 \pm 1°$ C due to structural transformations (see II-4.1). Norsethite, $BaMg(CO_3)_2$, transforms at the same low temperature (811° C), but at high transformation temperature, being 12° C lower than that of witherite. Witherites containing ~1 weight-% CaO+MgO invert at $978 \pm 0.4°$ C. Both the data of the second transformation temperature of this witherite containing Ca and Mg and of norsethite demonstrate that this second transformation temperature is apparently being influenced by the Mg contents incorporated into the witherite resp. norsethite structure.

Strontianites ($SrCO_3$), which should invert at 929° C from an orthorhombic low- to a hexagonal high-temperature modification (after D'ANS-LAX), in DTA curves mostly show several little endothermic peaks lying close together between 880 and 930° C (ΔT: 1–2° C), all being reversible (SMYKATZ-KLOSS, 1967b). A very pure strontianite sample from Drensteinfurt/Westfalia had only one single endothermic peak at 901° C, a yellowish strontianite from the type found in Loch Strontian/Scotland, characterized by a BaO-content of 0.4 weight-%, showed two endothermic peaks at 894 and 923° C, and a Ca-bearing strontianite from Ahlen/Westfalia could be recognized at peak temperatures of 893 and 917° C (exactness of measurement in all cases $\pm 0.5°$ C). These data demonstrate that the transformation temperatures of strontianites are largely influenced by the incorporation of Ca and Mg into the strontianite structure.

2.2 Cu-Ag Sulfides

Many chalcogenide minerals exist in high- and low-temperature modifications, the transformation temperatures between these high- and low-forms being very characteristic for the different minerals. Above all the sulfides, antimonides, selenides etc. of silver, copper, zinc and nickel can be recognized very well by such inversions (HILLER and PROBSTHAIN, 1956; BOLLIN; FRANZ). Especially the minerals of the system Cu-Ag-S have been studied in detail (POSNJAK et al.; N.W. BUERGER; BUERGER and BUERGER; KRACEK; SUHR; DJURLE; SKINNER; ROSEBOOM; SADANAGA and SUENO). But in the case of spontaneous displacive structural transformations (inversions) of these minerals you can find very different inversion temperatures ($= t_i$): 173–179° C for the inversion between monoclinic low-temperature Ag_2S (acanthite) and cubic high-temperature Ag_2S (argentite), 91° and 103–105° for the inversion between orthorhombic low- and hexagonal high-temperature chalcocite (stoichiometric Cu_2S), 78° and 94° for the inversion between orthorhombic and hexagonal strohmeyerite, $Cu_{1+x}Ag_{1-x}S$ ($\sim CuAgS$), 112° and 117° for the inversion between high- and low-jalpaite, $Cu_{0.45}Ag_{1.55}S$, and 76–83° for high-low digenite, $Cu_{1.78-1.83}S$. The minerals djurleite, $Cu_{1.96}S$, mcinstruyite, $Cu_{0.8}Ag_{1.2}S$ and covellite ($\sim CuS$) transform irreversibly when heated into mixtures of other phases (SKINNER), showing endothermic DTA peaks at 93° C (djurleite), 94–97° (mcinstruyite) and $507 \pm 3°$ C (covellite; all data for p = 1 atm).

The DTA curves of ten natural Cu-Ag-sulfide samples heated up to 300° C under standard conditions showed endothermic effects between 77° and 103.5° C resp. between 160.5° and 177.5° C. These were reversible in second and third runs and appeared as exothermic effects on cooling. Besides these reversible peaks another endothermic effect appeared in some curves between 88.5 and 92.5° C which was not reversible. The minerals listed in the following Table 28 were determined by additional X-ray, optical and chemical investigations. The samples were chemically analyzed only for the main components Cu and Fe in "chalcocites" and Ag in Ag_2S (average values of three analysis). From the acanthite of Cerro de Pasco, no material sufficient for chemical analysis was available. The temperature determinations were made with the aid of the two internal standards KNO_3, which shows a sharp endothermic effect reflecting the α–β inversion, and water-free Na_2SO_4 with two inversions at 147 and 245° C. By being calibrated with these internal standards, the exactness of measurement by DTA was $\pm 0.3°$ C. With the exception of the Cerro de Pasco-sample material, several DTA determinations were made from all samples.

Surprising is the high iron content of three samples. Only one of them contains finely dispersed hematite (Guilt "red") found by X-rays. On investigating seven "chalcocite" samples, it was found that all consist of at least two different copper sulfides, the main mineral of five samples being chalcocite, the main phase of the last two samples being djurleite, which occurs in 6 of 7 samples. Besides chalcocite and djurleite the min-

Table 28. Transformation temperatures of Cu-Ag sulfides (in °C, ± 0,3°)

Sample, origin	Weight-%			Transformation temperature				
	Ag	Cu	Fe	Ag_2S	Digenite	Djurleite	Unknown	Chalcocite
Acanthite, Cerro de Pasco, Peru	X			160.5 w		91 w		
Acanthite, Chanarcillo, Chile	85.2			177.5 ss				
Acanthite, St. Andreasberg, Harz (Germany)	88.6			177.5 ss				
Cu_2S, Tsumeb		81.8	1.5		85 cl	92.5 ss		
Cu_2S, Butte-II, Mont.		76.9	3.55		82 w	91.5 cl		103.5 ss
Cu_2S, Lowell, Arizona		72.6	4.8			90.8 w		102.5 ss
Cu_2S, Butte-I, Mont.		75.8	6.2			90.2 w		102.0 ss
Cu_2S, Guilt (red.), Neue Hardt, Germany		25.6	43,4+ (8)		77.5 w	89.5 s		
Cu_2S, Guilt (black), Neue Hardt	X	—	10 ± 1			88.7 s		99.5 s
Cu_2S, Pinal Co., Arizona		42.9	18.3				82 cl	95 ss

DTA deflections: w = weak, s = strong, cl = clear, ss = very strong.
X = too little material for an exact chemical analysis, both samples Fe-bearing; Guilt "black" ~ 10% Fe.
+ = impurity of fine-dispersed hematite (~ 35% Fe!) being the cause of red colour.

eral digenite occurs three times, most frequently in samples containing only minerals poor in Cu, the non-stoichiometric sulfides digenite and djurleite.

Two of the three Ag_2S-samples only consist of acanthite. The third sample contains acanthite and djurleite, the last one being the most frequently occurring sulfide mineral in all samples investigated. The fact that djurleite has not become well-known from sulfide deposits until recent years may be caused by its similarity to chalcocite in X-ray diagrams and chemical composition. By means of DTA, both minerals can be easily distinguished or their existence proved even in the case of coexistence (compare with Fig. 65).

The transformation temperatures of Cu-sulfides depend on the chemical composition, and mainly on the Fe-content being incorporated in

118 Part III. Special Application of Differential Thermal Analysis

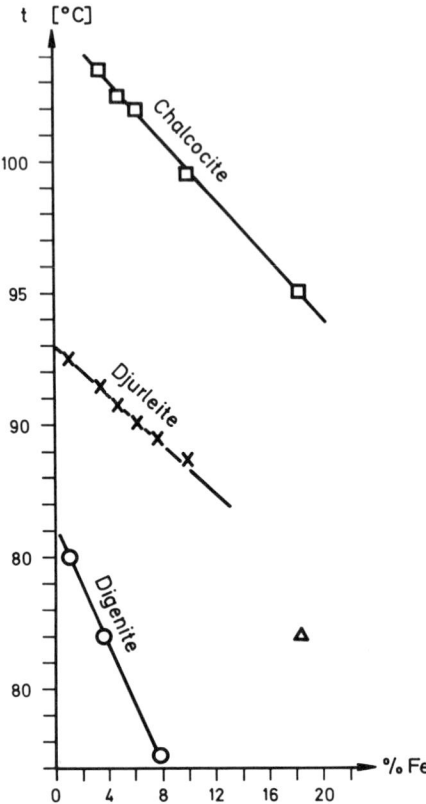

Fig. 65. Dependence of the transformation temperatures of chalcocite, djurleite, and digenite on Fe-content

the sulfide structure, so having formed solid solutions of copper sulfides with copper-iron sulfides. Lowering of the inversion temperature of chalcocite and digenite and of the irreversible transformation temperature of djurleite with increasing Fe-contents can be seen very clearly (Fig. 65). To what the distinct endothermic effect at 82° C in the DTA curve of the Fe-rich sample "Pinal" belongs is still obscure: the X-ray diagram of this sample only shows the interferences of chalcocite.

3. Influence of the Chemical Composition on the Curie-Temperatures of Magnetites

At the Curie-temperature, ferromagnetic substances like nickel or the mineral magnetite lose their ferromagnetism and become paramagnetic. This change of magnetic properties, which is very characteristic for each ferromagnetic material, can be recognized in a DTA curve as a very slight but still sharp endothermic peak (ΔT: 0.04–0.3° C). By using a nickel block sample holder or nickel crucibles, the Curie-temperature of nickel can be seen in the DTA curve at 360° C. The Curie-temperature of pure magnetite, Fe_3O_4, occurs at $580 \pm 2°$ C (SCHMIDT and VERMAAS; BLUM et al.). It will not be modified during the ensuing heating or cooling. Substitutions of Fe^{2+} and Fe^{3+} by Al^{3+}, Cr^{3+}, Ti^{4+}, Mg^{2+}, Ni^{2+}, Ca^{2+}, Mn^{2+} and others which are partly very common in natural magnetites *lower* the Curie-temperature Θ after VINCENT et al. from 580° for 100 weight-% Fe_3O_4 to 450° C for a magnetite consisting only of 80 Weight-% Fe_3O_4, to 300–350° C for 60 weight-% and down to 100–150° C for 40 weight-% Fe_3O_4 (the remainder being TiO_2, Cr_2O_3, MgO etc.). More suitable than a diagram of Curie-temperature versus Fe_3O_4-content (in weight-%) as VINCENT et al. have drawn it, will be the construction of a diagram of Curie-temperature versus TiO_2, Cr_2O_3 etc. (see below).

Among minerals, ferromagnetic substances are the (titano-) magnetites (Θ: 580° C), maghemite, pyrrhotite, some hematites, (a very weak change of magnetic properties reflected by a small endothermic DTA peak at 680° C) and titanohematites (Θ: 600–660° C, CHEVALLIER et al.). The sometimes reported ferromagnetism of some cassiterites had been caused by admixtures of magnetite or by inclusions of $Fe_{2-x}Ti_xO_3$ (Θ: 50–250° C, GRUBB and HANNAFORD). The measurement of *Néel-temperatures* (when antiferromagnetic substances lose their antiferromagnetism) is only possible in a low-temperature DTA apparatus, because Néel-temperatures generally occur below $-50°$ C. Néel-temperatures of the wolframite solid solutions ($FeWO_4$-$MnWO_4$) vary regularly with linearly increasing $FeWO_4$- or MnO_4-content, so giving a criterion for the determination of the wolframite composition (WEITZEL). Olivines, too, show typical antiferromagnetic properties (DUFF) and their loss (= Néel-temperature) depends on the Mn^{2+}-, Co^{2+}-, Ni^{2+}-contents of the olivines. The Curie- as well as the Néel-temperatures will generally be measured by other methods than DTA. But the following studies show that, contrary to the opinion of LEWIS, in case of magnetites DTA measurements of Curie-temperatures can help to classify and characterize natural magnetites very well and very simply. For this reason the Curie-temperatures of 23 natural magnetites, mostly from Tertiary basalts from North Germany or from Scandinavian iron ores, have been measured (standard conditions with the exception of the use of a nickel sample holder). All the Curie-temperatures of these 23 samples lie between 490 and $586 \pm 2°$ C.

As mentioned above, substitution processes are responsible for a lowering of Θ. Therefore in Fig. 66 all magnetites with known Curie-temperatures and known chemical composition were drawn in a dia-

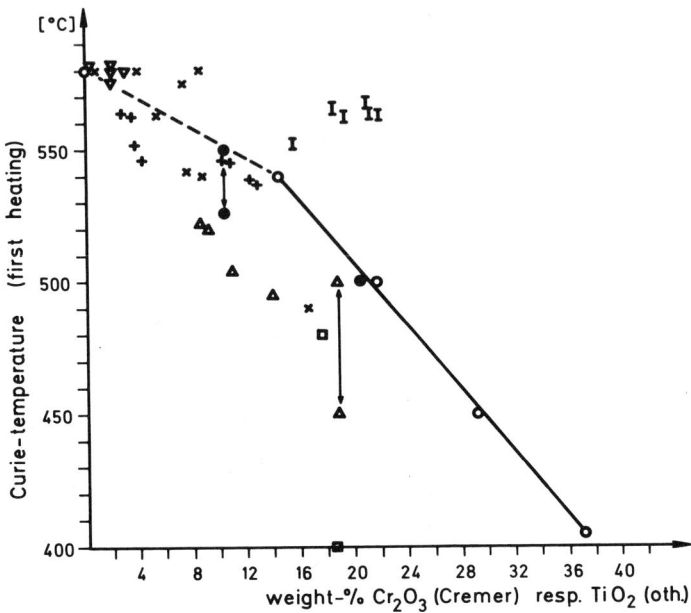

Fig. 66. Dependence of the Curie-temperatures Θ on the Cr_2O_3- resp. TiO_2-contents of magnetites. $\bigcirc$ = synthetic Ti-magnetites (CHEVALLIER et al.), ● = synthetic chromite-magnetite (CREMER), Δ = natural Ti-magnetites (CHEVALLIER et al.); + = natural Ti-magnetites (FOMINYKH and GLUKHIKH); x = the author's data; □ = natural Ti-magnetites (LEWIS); ▽ = natural magnetites (SCHMIDT and VERMAAS); I = natural magnetite-ulvite intergrowths (VINCENT et al.)

gram Θ versus Cr_2O_3- resp. TiO_2-contents. Among the data of seven publications (see legend to Fig. 66), only the data of SCHMIDT and VERMAAS and the author's own values were determined by means of DTA. The inked line of Fig. 66 is valid for synthetic magnetites (values from CHEVALLIER et al.) and for one chromite-magnetite mixed crystal (CREMER; mole-% converted in weight-%). The scattering of the drawn points of natural titano-magnetites, which partly exceeds a value of 100° C and from which LEWIS drew his conclusion that the Curie-temperatures of natural magnetites do not allow statements on their chemical compositions, can be explained very simply: all values lying to the left of the marked line will only give *one part* of the substituents incorporated into the magnetite structure. By supplementing the amount of TiO_2 (Cr_2O_3) by the contents of *all* those elements which substitute the iron, all values will come to lie on the inked line or very close to it,—with the exception of the two values given by LEWIS, the deviation of which is still obscure. The six values from VINCENT et al. to the right of the line (signature: I) represent natural *intergrowths* of magnetite and ulvite (Fe_2TiO_4). This means that the high TiO_2-contents (15–22 weight-%) of these samples are *not* incorporated into the magnetite structure (and only in the case of substituents being incorporated in the magnetite structure would the Ti-content influence the Curie-temperature!). In this case the chemical analysis simulates an impure magnetite, but really the magnetite structure contains no or only very few TiO_2. The corrected values of VINCENT et al. (those are the given values of Fig. 66 substracted by the TiO_2-contents) all fall on the broken line in Fig. 66. This correction is valid, as can be seen in the thermal behaviour (tempering) of these intergrowths. VINCENT et al. obtained a homogenization of the intergrowths by heating the samples up to 950° C: at this temperature only one phase had been recrystallized. During the slow heating process VINCENT et al. observed a continuous lowering of the Curie-temperature until Θ reached 450–500° C for the completely homogenized phase, a titanomagnetite having incorporated 20–22 weight-% TiO_2 into its structure.

Figure 67 contains only the values of synthetic magnetites and those of natural samples, being true magnetites and not intergrowths. All samples used in the diagram of Fig. 67 are well-known in their chemical composition (100 ± 0.4 weight-%). From these values result two lines for synthetic, respectively for natural samples not totally identical. For differential thermal analytic determinations of the Curie-temperatures of natural (titano-) magnetites the inked line of Fig. 67 should be used. Such a determination of Θ of natural titanomagnetites by means of DTA combined with a chemical determination of TiO_2 can by the aid of Fig. 67 lead to the following conclusions:

a) A Θ value lying to the right of the inked line indicates that the titanium is only partly incorporated into the magnetite structure. With a margin of error of two weight-% TiO_2 it is possible to estimate the part of titanium substituent in the magnetite structure and the remaining part that is a separate Ti-phase intergrown with the magnetite.

b) A value lying to the left of the inked line indicates the sum of the contents of substituents others than TiO_2.

c) By heating a sample consisting of intergrowths of magnetite and ulvite (or ilmenite, or rutile) up to 1000° C the sample will become homogenized, forming a titanomagnetite. The lowering of the Curie-temperature will be a measure for the degree of this homogenization.

This method will be highly suitable in cases of submicroscopical lamellae of ilmenite in magnetites. Further the method can be applied for the petrographic classification of basaltic rocks by means of investigation of their (titano-) magnetites as has been described by KOHN or WRIGHT, who took the chemical composition of magnetites to point out petrogenetic developments.

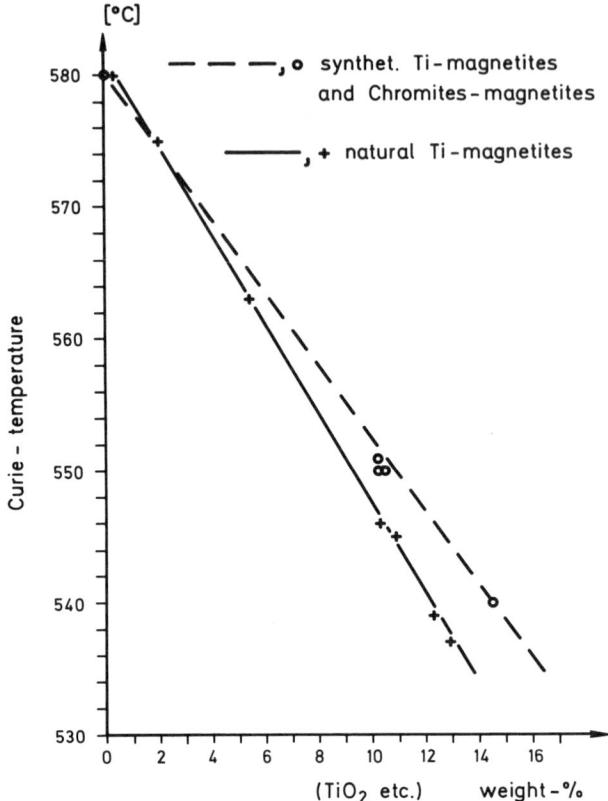

Fig. 67. Curie-temperatures versus the contents of $Fe^{2+,3+}$-substituents of some magnetites

The investigation of one *maghemite* sample (more samples of the mineral were not available for this study) showed a Θ-value of $640 \pm 5°$ C for this cubic Fe_2O_3, the value lying just between Θ of pure magnetites (580° C) and Θ weakly appearing in the DTA curves of some hematites (680° C). Thus it seems possible to distinguish differential thermal analytically between both the cubic forms of iron oxides, magnetite and maghemite.

4. Contribution to the Classification of Chlorites

The sheet silicate minerals of the chlorite group (compare with II-7.5) are generally subdivided according to their optical properties, which reflect their chemical composition (see TROCHIM in TRÖGER-BRAITSCH:

proposals to the classification of chlorites from TSCHERMAK, ORCEL, WINCHELL, HEY, BRINDLEY, SCHÜLLER a.o.). ALBEE distinguished the minerals of the four main subgroups Mg-chlorites (optical positive, normal interference colours), Mg-Fe-chlorites: (optical positive, anomal brown interference colours), Fe-Mg-chlorites: (optical negative, anomal blue interference colours) and Fe-chlorites: (optical negative, normal interference colours). But this scheme is incomplete and not valid for all possible Al-Si-distributions in chlorites (TROCHIM). After TROCHIM the optical determination and classification of chlorites will not be successful to the same extent as in optical investigations of other mineral groups-, "in spite of the optimistical estimation of some authors" (TROCHIM). Therefore as a method additional to optics TROCHIM recommends "the differential thermal analysis and the possibilities of classification latent in this method".

The extremely fine grain size of most chlorites ($<2\,\mu\,\varnothing$) often makes an optical investigation impossible, above all in sedimentary chlorites. Therefore it has often been attempted to find out the relation between the thermal behaviour and the chemical composition of chlorites by heating series of chlorites formed hydrothermally and well-known in their chemical composition (CAILLÈRE and HÉNIN; PHILIPPS; LAPHAM). The results of these authors cannot easily be compared because of great differences in the conditions of analyses. Therefore the author's analysis was made under highly standardized conditions of analysis (Table 1, but heating in a nickel block sample holder after the samples had been ground for three hours in an agate ball mill) using only the grain size fraction $0.6\text{--}2\,\mu\,\varnothing$. Only in one sample (see Table 19, II-7.5) were the other fractions analyzed, too, the purpose being the possibility of estimating the influence of grain size on the DTA data of chlorites. Fig. 43 (II-7.5) contains some DTA curves of chlorites with increasing Mg-contents from top to bottom. The substitution of the octahedral central ion Mg^{++} (see II-7.5) by Fe^{++}, Al^{+++}, Mn^{++}, or Cr^{+++} is reflected in the DTA curves: increasing substitution of Mg^{++} by these ions (with exception of Cr^{+++}, see below) shifts the decomposition peaks of chlorites down to lower temperatures (see Figs. 43 and 68). However, the influence of Mn^{++} on the lowering of the decomposition temperature is very slight. Cr^{+++} at the position of Mg^{++} in the center of the octahedral layers leads to an *increase* of the structural chlorite stability (in opposition to all other substituents by which the structural stability was *decreased*) which can be explained by crystal chemical facts (see II-7.5). Besides this possibility of being incorporated into the octahedral layers, which is the case in the violet-coloured Cr-chlorite kaemmererite, the chromium can also be fourfold coordinated. It is incorporated into the tetrahedral layers, then substituting Si^{4+} or Al^{3+} in the case of the red-coloured Cr-chlorite kotschubeite. The author's DTA data of two chemi-

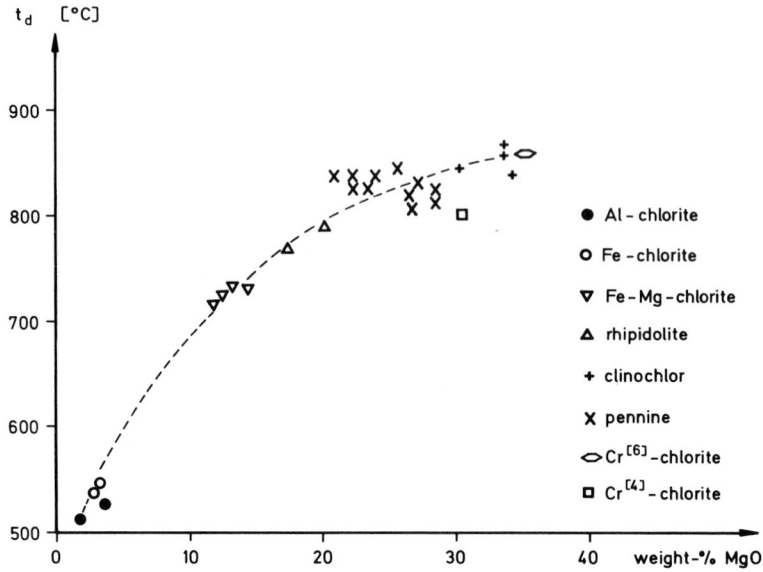

Fig. 68. Dependence of decomposition temperatures on MgO-contents in chlorites

cally analyzed Cr-chlorites corroborate the results of LAPHAM: kaemmererite (with $Cr^{[6]}$), is decomposed at high temperatures and kotschubeite (with $Cr^{[4]}$) at lower temperatures than could be expected according to the Mg-content of this last mineral (compare with Fig. 68).

A correlation between the peak temperatures of the first endothermic deflection, caused by release of OH^- and by decomposition of the Fe-Al-component of the octahedral layers and the Fe-contents of these chlorites determined by the chemical analysis, gives no clear relationship. But the diagram showing MgO-contents versus decomposition temperatures of the chlorites (= the second endothermic deflection in curves of Mg- and of Mg-Fe-chlorites or the only endothermic deflection in curves of Fe- and of Fe-Mg-chlorites) shows an excellent correlation line (Fig. 68) highly suitable for characterization and classification of chloritic minerals. But this diagram of Fig. 68 will only be valid for chlorites from igneous or metamorphic rocks, that means for chlorites that are all well-ordered, and for minerals of the grain size fraction 0.6–2 μ ∅. If the diagram should be valid for sedimentary chlorites, too, some additional "simplifications" have to be made (see below). All samples listed in Table 19 (II-7.5) and Fig. 68 have been analyzed chemically for their MgO-, $(FeO+Fe_2O_3)$-, MnO-contents, and ten samples have even been fully analyzed chemically.

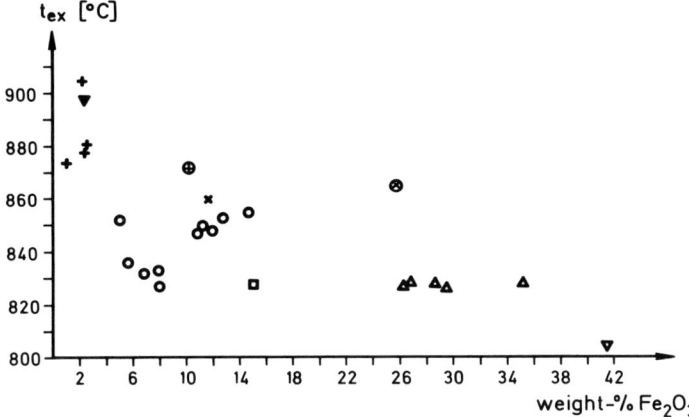

Fig. 69. Dependence of the temperature of the exothermic peak on the Fe_2O_3-content of chlorites

The *exothermic* peak appearing in DTA curves of most chlorites between 800 and 900° C also depends in its peak temperature on the chemical composition of the chlorites. This exothermic peak caused by the formation of olivine and spinel shows a distinct dependence on the Fe_2O_3-content and a lesser dependence on the MgO-content of the chlorites; of course, in DTA curves of pure Fe-chlorites (thuringites, chamosites) the exothermic peak will very rarely be observed.

From Figs. 68 and 69 all non-sedimentary chlorites can be put into one of the eight "subclasses" listed in Table 29. Ni^{++}-chlorites are to be treated like Mg-chlorites, since there is no possibility of distinguishing them because of the same crystal chemical behaviour (e.g. exactly the same ionic radius: 0.78 Å). The (slight) influence of the divalent manganese substituting Mg^{++} on the thermal behaviour of chlorites cannot be measured, either, the reason still not being clear.

The *serpentine* minerals, the thermal behaviour of which is generally very similar to that of chlorites (see II-7.6), can nevertheless be distinguished clearly from chlorites, as can be seen in Table 30.

Sedimentary chlorites from soils and sediments are mostly mixed very deeply with other clay minerals or (hydr-)oxides and cannot be separated quantitatively. Therefore it is very difficult to obtain a chemical analysis of sedimentary chlorites. However, two samples impurified very slightly by other minerals ("earthy chlorite from Hohenstein, Saxony" and "Al-122", see Table 19) gave decomposition temperatures of 720° C (chlorite from Hohenstein; 28.3 weight-% MgO; 13.25% Fe_2O_3) and 617° C, ("Al-122"; 11.2% MgO; 29.4% Fe_2O_3), the temperatures of the exo-

Table 29. Interdependence between chemical composition, decomposition temperatures (t_d), and temperature of the exothermic peak (t_{ex}) of chlorites

Sample	Weight -% MgO	Weight -% Fe_2O_3 (tot. Fe)	t_d (°C)	t_{ex} (°C)	a-Value (°C)	Class
Leuchtenbergite, Ural, USSR	34.2	2.25	837	—	—	clinochlor
Leuchtenbergite, Kärnten/Austria	33.6	2.54	864	881	17	
Leuchtenbergite, Sunk/Trieben/Austr.	33.5	2.4	856	878	22	
Leuchtenbergite, Neuberg/Austria	30.1	1.05	845	874	29	
Grochauite, Zillertal, Alps	25.7	10.2	847	872	25	
Kaemmererite (8.7 wgt.-% Cr_2O_3)	35.12	—	858	—	—	$Cr^{[6]}$-chlorite
Pennine, Rimpfischwäng, Switz.	28.4	5.6	820	836	16	pennine
Pennine, Rimpfischwäng, Switz.	28.4	6.75	818	832	14	
Pennine, Rimpfischwäng, Switz.	26.6	8.0	809	827	18	
Pennine, Rimpfischwäng, Switz.	26.45	7.9	817	833	16	
Pennine, Marktredwitz, Germany	27.1	5.0	832	852	20	
Pennine, Marktredwitz, Germany	23.9	11.15	833	850	17	
Pennine, Marktredwitz, Germany	23.5	11.95	827	848	21	
Pennine, Marktredwitz, Germany	22.6	12.7	833	853	20	
Pennine, Marktredwitz, Germany	22.55	10.8	827	847	20	
Pennine, Marktredwitz, Germany	21.3	14.7	833	855	22	
Kotschubeite (5% Cr_2O_3)	30.4	14.9	801	828	27	$Cr^{[4]}$-chlorite
Prochlorite, Chaffee Co., USA	20.25	11.6	791	860	69	ripidolite
Prochlorite, Kaareck, Austria	17.4	25.85	773	865	92	
Pseudothuringite, Maltatal	14.35	28.6	731	828	97	Fe-Mg-chlorite
Aphrosiderite, Al-286	13.12	26.15	730	827	97	
Aphrosiderite, Al-288	12.8	26.7	725	828	103	
Aphrosiderite, Nassau, Germany	12.5	25.6	721	828	107	
Thuringite, Schmiedefeld, Germany	2.75	49.7	539	—	—	Fe-chlorite
Fe-chlorite, Canaglia, Sardinia	2.84	41.5	541	805	264	
Lias-chlorite, CSSR	3.7	9.5	526	—	—	Al-chlorite
Cookeite, Cornberg, Germany	1.6	2.3	511	898	387	

Samples kindly supplied by:
Drs. F. J. Eckardt and P. Müller, Bundesanstalt f. Bodenforschung, Hannover;
Prof. V. Kupčik, Göttingen;
Prof. Germ. Müller, Heidelberg;
Prof. H. Meixner, Salzburg, Austria;
Dr. W. Echle, Aachen.

thermic peak lying at 823° C (chlorite from Hohenstein) resp. at 826° C ("Al-122"). According to the chemical composition the sample from Hohenstein is a pennine, "Al-122" a Fe-Mg-chlorite. While both chlorites were decomposed at nearly 100–120° C lower than non-sedimen-

Table 30. Comparison of DTA data of chlorites and serpentines

Class	t_d(°C)	t_{ex}(°C)	a-Value(°C)
Clinochlor	837—864	874—881	17— 29
$Cr^{[6]}$-chlorite	858	—	—
Pennine	809—833	827—855	14— 22
$Cr^{[4]}$-chlorite	801	828	27
Ripidolite	773—791	860—865	69— 92
Fe-Mg-chlorite	721—731	827—828	97—107
Fe-chlorite	520—580	800—810	264
Al-chlorite	500—530	895—900	387
Chrysotile	750—780	820—825 (ΔT very large)	48
Antigorite	720	835	115

tary pennines resp. Fe-Mg-chlorites, the temperatures of the *exo*thermic peak do *not* differ from those of igneous or magmatic rocks that are chemically similar. The fact that the greater *degree of disorder* of sedimentary chlorites is the reason for the lowering of the structural stability is reflected in these lower decomposition temperatures. It will, however, have no influence on the formation of new phases (olivine, spinel) that are reflected in the exothermic DTA effect. This exothermic effect will therefore be a suitable measure for the chemical composition of hydrothermally or metamorphically formed chlorites and for sedimentary ones as well, because this effect is not influenced by crystal chemical properties of the samples like, for instance, disorder phenomena.

In DTA curves of more than a hundred chlorite-bearing clays (SMYKATZ-KLOSS, 1966), the decomposition temperatures of these sedimentary chlorites were between 605 and 690° C. Having been corrected by this "value of experience" (= 100–120° C lower than in the case of non-sedimentary chlorites), all these chlorites should belong to the subclasses ripidolite, clinochlor and pennine. The classification on account of the exothermic peak temperature may be more exact, but the exothermic effect cannot be observed in all cases: it is absent in DTA curves of iron-rich clorites. This chemical variety rarely occurs in soils and sediments. The exothermic peak cannot be observed in very strongly disordered chlorites which are characterized by decomposition temperatures lying more than 150° C below those of non-sedimentary samples of comparable chemical composition. This last type of strongly disordered chlorites, however, can easily be recognized by very broad DTA deflections with very low ΔT-values.

As one chlorite sample described by CHEN is a Cr-bearing thuringite (0.22 weight-% Cr_2O_3; 38,5% $FeO + Fe_2O_3$; 5.25% MgO) showing a strong decomposition peak at 590° C, it corresponds very well with the data reported in this study.

5. Smectites and Vermiculites: The Distinction between Di- and Tri-Octahedral Minerals and Grain Size Determination

The minerals of the montmorillonite group (smectites) can be distinguished from the vermiculites, as they are very similar to them in their thermal behaviour, as regards the temperatures of the exothermic peak. This peak appears in DTA curves of silicates rich in Al (= kaolinites, Al-chlorites, montmorillonites) at >900° C. Smectites are mostly di-octahedral (Al-rich montmorillonites); the tri-octahedral types hectorite, saponite and sauconite occur very rarely, whereas vermiculites are generally tri-octahedral. They can, of course, become partly di-octahedral by substitution of the main octahedral cation Mg^{++} by Al^{3+} or Fe^{3+}. The temperature of the exothermic peak is a suitable measure for the di-octahedral portion of a vermiculite, and further a hint at the nature of the substituent (Fe^{3+} or Al^{3+}): the exothermic peak temperature of a "normal" tri-octahedral Mg-vermiculite lies between 830° and 850° C (GRIM and ROWLAND; BARSHAD). With increasing substitution of Mg^{++} this temperature will be lowered in case of Fe^{3+} being the substituent, or it will be raised in the case of Al^{3+} being the substituent. A (hypothetic!) di-octahedral vermiculite that has occupied half of the central positions of the octahedral layers by Fe^{3+} and half of them by Al^{3+}, however, would *not* differ in its exothermic peak temperature from that of a pure Mg-vermiculite. But in all naturally occurring vermiculites not more than half of the Mg^{++} will be substituted, and in all cases of substitutions only *one* trivalent cation as the predominant substituent, Al^{3+} *or* Fe^{3+} (e.g. compare with JASMUND), can be observed. According to the exothermic peak temperature two of the four vermiculites listed in Table 17 (II-7.3) are "normal" Mg-vermiculites, and the two last samples of Table 17 show how a distinct portion of di-octahedral Al-vermiculite is reflected in a shifting of the temperature of the exothermic peak up to 865° resp. 878° C. This possibility of estimating the amount of substitution of Mg^{++} by Al^{3+} or Fe^{3+} in vermiculites is to be considered as a preliminary method; the number of vermiculite samples investigated was too small for putting up a calibration curve.

Table 31. t_d and t_{ex} of several smectites

Sample, origin	Type	Octahedral cations (predom.)	t_d (°C) Fe-, Al-, component	Mg-	t_{ex} (°C)
Montmorillonitic clay, Hesse	di-octahedral	Al^{+++}	725		947
Montmorillonite, Upton, Wyo.	di-octahedral	Al^{+++}	702		936
Montmorillonite, OECD	*di*-tri-octahedral	Al^{+++}, Mg^{++}	674	880	1000
Otayite, Calif.	*tri*-di-octahedral	Mg^{++}, Al^{+++}	653	855	986
Hectorite, Calif.	tri-octahedral	Mg^{++}		769	1010
Saponite, Fichtelgebirge, Germany	tri-octahedral	Mg^{++}, Fe^{++}		842	985
Nontronite, Harz, Germany	di-octahedral	Fe^{+++}	470		950

In the case of smectites where "pure" di- as well as tri-octahedral types occur and where they were available for the present study two criteria are found for a clear distinction between di- and tri-octahedral types. The first is the temperature of the decomposition peak ($= t_d$), and the second the temperature of the exothermic peak ($= t_{ex}$). For all di-octahedral smectites t_d and t_{ex} are very much lower than for tri-octahedral types (compare with Table 17, II-7.3). Besides "pure" di- or tri-octahedral minerals, mixed types that are *di*-tri-octahedral (e.g. the OECD sample, see Table 31) or *tri*-di-octahedral like "otayite" (Table 31), may occur. The "otayite", especially, is very rare: it has been described as being a di-octahedral mineral ("montmorillonite"), as well as being a tri-octahedral type ("otayite"). According to its chemical composition (compare with JASMUND) it belongs to the smectites, being rich in MgO ($=$ tri-octahedral) showing MgO-contents of 6.5 weight-%. The montmorillonite from Upton, Wyoming, on the other hand, is certainly a di-octahedral smectite type (2.5% MgO). According to its thermal behaviour as well as its chemical composition, this type from Otay should not be marked as "montmorillonite", but more clearly as *otayite*.

The *nature* of the *interlayer cation* (Mg^{++}, Na^+, H_3O^+, Ca^{++}, K^+) that is necessary to compensate the negatively charged packets of tetrahedral-octahedral-tetrahedral layers of smectites and vermiculites can possibly be seen from combined rehydration and differential thermal investigations: The endothermic deflection between 105° and 200° C reflects the dehydration of the interlayer spaces and seems to be dependent in its peak temperature on the type of the predominant interlayer cation. More detailed investigations on this problem have been started.

Table 32. Width at half height of dehydration peaks of montmorillonite and halloysite in dependence on grain size

Grain size fraction	Width at half height (°C)	
	montmorillonite	Halloysite
$> 2\,\mu\,\varnothing$	34	—
$2-0.6\,\mu\,\varnothing$	45, 48	49
$< 0.6\,\mu\,\varnothing$	52	58

After the dehydration of smectites, vermiculites or halloysites and a following re-hydration with water (at comparable degrees of humidity), the first endothermic deflection (peak between 100° and 200° C) can be taken as a measure for the *grain-size* of these clay minerals. This could be seen in DTA runs of various grain size fractions of montmorillonite and halloysite samples (Table 32). Standard conditions of analysis were employed (but Ni-block sample holder); in all runs the same degree of humidity (not measured); rehydration time: 2 hrs, then drying at 60° C for 2 hrs.

A suitable measure for the grain size of hydrous clay minerals is the width at half height of the dehydration peak (see Table 32), in °C at $\Delta T/2$. The re-hydrated samples show only one endothermic peak below 200° C, in contrast to most "fresh" montmorillonites, which show two or more separate endothermic effects in this temperature range due to the dehydration of the adsorbed and the loss of the interlayer water. The shape of this endothermic deflection is well-rounded and shows no distinctly marked top; ΔT decreases slightly, but the width at $\Delta T/2$ increases with decreasing grain size. Montmorillonites and other clay minerals become more active with decreasing grain size (EARLEY et al.), the reason being a larger surface which can adsorb more water than the smaller surface of a coarser sample with the same weight. The differences of the widths at half height of dehydration peaks of various grain size fractions seem to be caused by different amounts of water being adsorbed and, indirectly, by different grain sizes. So from the measurement of these widths at $\Delta T/2$ of the dehydration peaks of rehydrated clay minerals, the grain size of these clay minerals can be estimated.

6. Determination of the Degree of Disorder in Kaolinites

Kaolinites from soils and recent sediments are frequently disordered: the different sheet packets were distorted in the direction of the crystallographic b-axis by an amount $n \cdot b_0/3$. That can be observed in X-ray

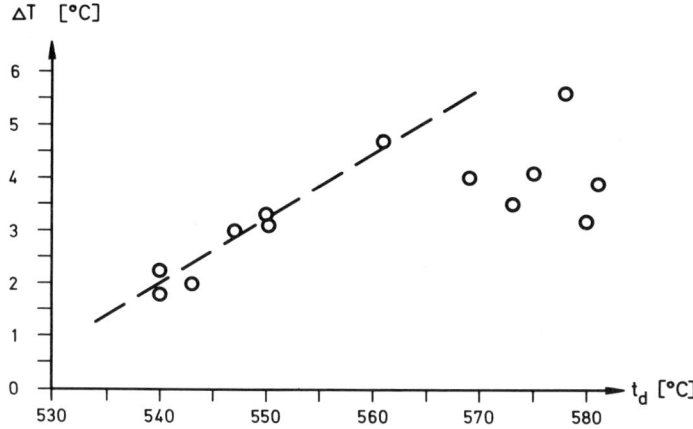

Fig. 70. Temperature versus ΔT of the endothermic (decomposition) peak of kaolinites

diagrams by an increasing diffusion of the six lines lying between the basal interferences of (001) and (002) to a broad band and by a broadening of the interferences for d-values <2 Å (BRINDLEY, 1961). During diagenetic processes the disorder of kaolinites decreases. So kaolinites from clay stones ("Tonsteine") from the Carboniferous of the Ruhr, Western Germany, completely lost their disorder during carbonization of the coal formation lying quite near (ECKARDT, 1963, temperatures ~200° C). All transitions between well-ordered and extremely disordered kaolinites can be found, as can be seen by means of X-ray, (continuous broadening of (0k0) interferences with increasing disorder) and of DTA as well (e.g. compare with VON ENGELHARDT and GOLDSCHMIDT: "fireclay-mineral from Franterre").

As shown in Fig. 70, not only the endothermic peak temperature of kaolinites, but also the ΔT-values of this peak decrease with increasing disorder. To a certain degree the temperature of the exothermic kaolinite peak also reflects the degree of disorder (see Table 15, II-7.1): the exothermic peak appears between 980° and 1005° C in DTA curves of well-ordered kaolinites, dickite and nacrite and decreases down to 940° ± 1° C (fireclay-mineral from Franterre) with increasing disorder.

The deviations from the broken line in Fig. 70 for well-ordered kaolinites can be explained by differences of grain size: all "fireclay"-samples (decomposition temperature $t_d < 555°$ C) and the investigated kaolinites from Murfreesboro ($t_d = 562$ and 578° C) belong to the grain size fraction 0.6−2 μ ∅. The five other samples characterized by relatively high decomposition temperatures, however, were slightly coarser (the maximum in their grain size distribution curves determined by the pipette-method lies in the fraction 2–6.3 μm ∅). The correlation between the DTA data and the widths at half height of the (hk0) X-ray interferences was impossible because of the broad band between $d=4.7$ Å and $d=3.4$ Å (see above), and because the intensities of the interferences <2 Å inclusive the (060) interference were too slight. But by this broad band all six samples that were decomposed below 555° C (samples from Franterre and from Großalmerode, Hesse) could be identified as being disordered.

The only difficulty in determining the degree of disorder by means of standardized DTA, and in achieving the calibration curve shown in Fig. 70, is the influence of grain size on ΔT. The diagram ΔT versus decomposition peak temperature (Fig. 70) is only valid for the grain size fraction 0.6–2 μm ∅. Kaolinites of other grain size differ in their ΔT-values. Actually a clearly defined correlation between grain size and ΔT cannot be determined (for the decomposition peak of kaolinites), but as the kaolinites are coarser than 2 μm ∅, they often show ΔT-values lower than those which could be expected from Fig. 70 according to their decomposition temperatures. The reason seems to be that the grain size *distribution* is more inhomogeneous in coarser- than in finer-grained samples. The ΔT-values increase with increasing homogeneity of the samples, but generally decrease with decreasing grain size (even if only to a small extent and even if this will be valid only for very fine-grained samples like clay minerals), so ΔT-values will be produced by two reasons effecting each other. The finer a sample is, the more homogeneous in its grain size it will be. But: it may become coarser in proportion to how well-ordered a kaolinite is.

A shifting of the peak temperature of the decomposition peak with varying grain size as described by BERKELHAMER et al. (1945) could not be observed in the present study. So the kaolinite, being the coarsest one, was decomposed at 580° C (the sample from Mesa Alta) and the two kaolinites from Murfreesboro, which were the most fine-grained samples studied, showed decomposition temperatures of 562° and 578° C.

The following consideration will be how to solve the problem of grain size influencing ΔT. The only interesting kaolinite samples to be studied by this method were the disordered, not the well-ordered types. Fortunately nearly all disordered kaolinites are very fine grained ($<2 \mu$ ∅) as well as very homogeneous in their grain size distribution. For this reason the diagram of Fig. 70 may really be used for the determination of the degree of disorder of kaolinites, provided that this determination is carried out under standard conditions of analysis.

The decomposition peak temperature alone will allow a first coarse classification into the following subgroups:

extremely disordered $(t_d < 530°\ C)$
strongly disordered $(t_d = 530–555°\ C)$
little disordered $(t_d = 555–575°\ C)$
well-ordered $(t_d > 575°\ C)$.

The t_d-value of 530° C seems to be very low according to all published DTA data of kaolinites. There ought to be only very few naturally occurring samples with decomposition temperatures below this value. Therefore it may be justified to take this temperature as "zero-value" for the exact determination of the degree of disorder:

1. The multiplication of the factor $(t_d–530)$ by the ΔT-value delivers a first index for this degree of (dis-)order (compare with SMYKATZ-KLOSS, 1974).

2. The relation of all indices to the index of well-ordered reference material (with the relative order = 100) gives the degree of order of each studied sample in per cent of the order of the reference kaolinite. The values of all investigated kaolinites (see Table 15) lie between 68% and 8% related to the reference kaolinite from Murfreesboro, which has been chosen to be the reference material because it is as fine-grained and homogeneous as the strongly disordered samples.

With the exception of the (hydrothermal) 2 M-type dickite all studied samples exhibit a lower degree of order than the reference kaolinite from Murfreesboro. But the degree of order of the kaolinites from Mesa Alta, South Carolina and Macon (see Table 15) obtained by this method seems to be too low because of the influence of grain size on the ΔT-values, as discussed above. The method developed can be used for the determination of the (dis-)order of kaolinites if the analyses were made under the proposed standard conditions (Table 1), and if a grain size fractionation (by means of the Atterberg method) is made delivering the fraction 0.6–2 µm ∅ (further discussion see SMYKATZ-KLOSS, 1974).

7. The Interdependence of Degree of Disorder, High-Low Inversion, and Temperature of Formation of Low-Temperature Cristobalites

The mineral cristobalite, being the high-temperature modification of SiO_2 (stable $>1470°$ C according to FENNER), occurs in a cubic high- and a tetragonal low-temperature form, both standing thermodynamically in equilibrium at 270° C (for p = 1 atm). At this temperature a displacive structural transformation (an inversion) takes place, which can be observed in DTA curves on heating (endothermic) and also on cooling (exothermic). But this temperature of inversion ($= t_i$) of 270° C is only valid for very well-ordered cristobalites (compare with SOSMAN), as they can be obtained by crystallization from a melt at ~ 1500° C (SABATIER, 1957), for instance. *Naturally*-formed cristobalites are always *disordered* in such a manner that only the strict layering of SiO_4-tetrahedrons one upon another will be disturbed one-dimensionally, being reflected in a lowering of t_i down to 60–80° C (FLÖRKE, 1955). The *term disorder* will include the sum of all structural defects, vacant positions, substitutions, disturbed surface etc. Differential thermal analysis is mainly a method very suitable for the recognition of the influence of disorder on the inversion behaviour of crystals (FLÖRKE, 1955; HOCHSTRASSER and FEITKNECHT, a. o.): Firstly increasing structural disorder will cause a lowering of inversion *temperatures* which reflects a decrease in structural stability, and secondly the *shape (form)* of a peak reflecting a structural inversion will be broadened by increasing disorder. In disordered parts of a crystal a structural change will begin a little earlier than in better-ordered parts (FLÖRKE, 1955). The inversion *temperature* will therefore become an inversion temperature *range*, a lot of discrete inversion temperatures which will give a very diffuse inversion effect over a temperature range of 50–100° C (see SMYKATZ-KLOSS, 1972a). That means: the relatively broad inversion effect summarizes all the discrete inversions taking place in various parts of different (dis-)order of a crystal. This is generally observed in DTA curves of cristobalites that have formed metastably in nature and are characterized by very small ΔT-values compared with the inversion of well-ordered synthetic cristobalite obtained at very high temperatures (SABATIER, 1957).

Cristobalite occurs metastably formed in pores of eruptive rocks (CORRENS), as a component of many agates and jaspers (RAMDOHR and STRUNZ; SMYKATZ-KLOSS, 1972a), in partly altered opals and even in soils, as for instance in the well-known Wyoming-bentonites (RAMDOHR and STRUNZ). The association of such "low-temperature cristobalite" (FLÖRKE, 1955) or "opal-cristobalite" (BRAITSCH) with chalcedony or opal refers to a metastable formation close to the surface at low temperatures. According to FLÖRKE (1962) strong structural disorder of minerals generally indicates very low temperature of formation. However, it may also be obtained by very quickly quenching a melt (compare with SMYKATZ-KLOSS, 1972a). The investigation of more than 50 microcrystalline SiO_2-samples by means of X-ray as well as DTA (see III-9) showed 17 samples containing disordered low-temperature cristobalite that were mostly accompanied by microcrystalline quartz (chalcedony). In two samples cristobalite was the main component. Data of Table 33 were obtained by standard conditions of analysis. Besides cristobalite and quartz in five samples, very small amounts of *low-tridymite* were located (to be recognized in DTA curves at t_i around 100 and 150–

Table 33. Inversion data of low-temperature cristobalites from agates, jaspers and one kascholong; I, II: two runs of the same rock sample

Sample	$t_i(\pm 2°C)$ I	II	Intensity (only qualitative)
12a, Violet core of jasper	80		Broad, very clear
12b, Bleached rim of jasper	183	183	Broad, middle-strong
13a, Light brown ball jasper	184		Broad, very clear
13b, Red core of ball jasper	185	177	Clear
15, Kascholong	208		Broad, very strong
16, Blueish-grey agate	190, 230	190, 230	Clear to strong
18a, Dense rim of agate	197		Clear
18b, Agate, red-white bands	167		Clear
18c, Agate, dark red and white bands	163	165	Very clear
18e, Agate, core, coarse-grained quartz	175		Middle-strong
20, Ivory coloured jasper	220, 245	206, 251	Strong
22a, Banded white jasper	190		Broad, very clear
22b, banded red jasper	188	195	Middle-strong
24b, Light brown core of jasper	177	167	Very clear to strong
25, Basalt jasper	97	93	Middle-strong
26a, White rim of jasper	94		Strong, sharp
29b, Light brown core of jasper	180		middle-strong

160° C, see II-3; not listed in Table 33). I and II: two DTA runs of the same sample.

The differences up to 14° C between the t_i obtained in two DTA runs of the same cristobalite rock sample indicate that there is a different degree of disorder of cristobalite crystals even in the same rock sample. The rounded shape of the endothermic inversion deflection (see Figs. 71 and 72), as well as the small ΔT-values of this peak, the temperature of which can only be measured with a margin of error of $\pm 2°$ C, show that all investigated cristobalites are more or less disordered (t_i between 80° and 245° C, see Table 33).

By hydrothermal hydrolysis of silicium nitride, cristobalite with different degrees of disorder was also formed metastably in the stability range of quartz (SMYKATZ-KLOSS and SCHULTZ). The synthesis conditions were 2 K bars P_{H_2O}, temperatures between 200° and 610° C, and run durations of two days. The application of DTA to investigate order-disorder phenomena in the newly formed phases turned out to be the most sensitive method to disorder effects.

From the DTA curves of Fig. 72 the degree of disorder of these cristobalites formed synthetically can be recognized by four characteristic features (as will also be the case in naturally formed cristobalites):

1. the temperature of inversion (t_i) from tetragonal (low) to cubic (high) cristobalite;

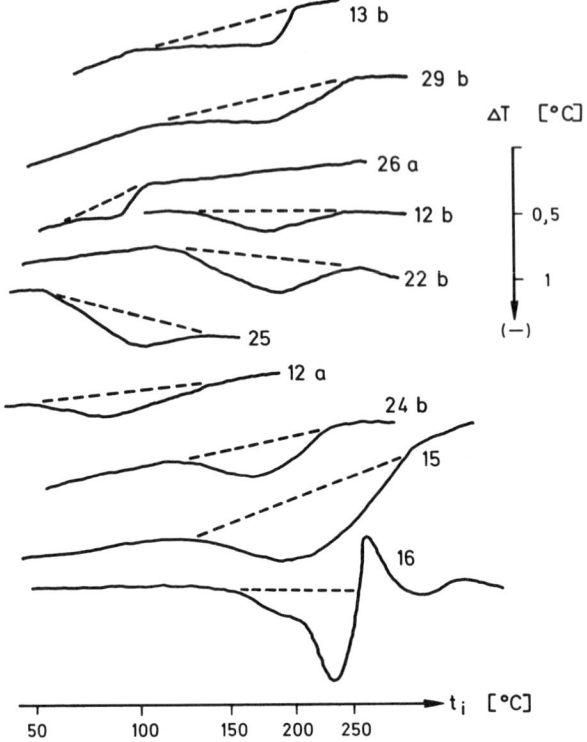

Fig. 71. DTA curves up to 300° C of cristobalites from micro-crystalline SiO_2-samples (agates, jaspers), first heating. Samples 15 and 22 b contain some opal, too (dehydration peaks < 130° C)

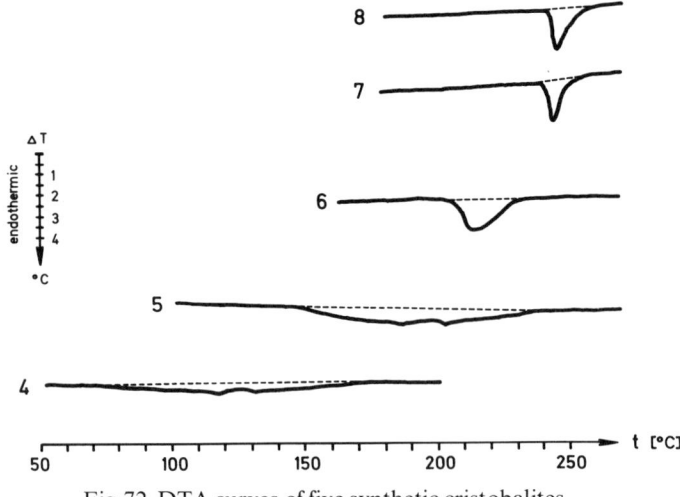

Fig. 72. DTA curves of five synthetic cristobalites

Table 34. DTA characteristics of synthetic cristobalites (p = 2 Kbar)

Sample (runs No.)	Temperature of synthesis (°C)	Appearance of peak in DTA curve	t_i (°C), first heating	t_i (°C), first cooling	Temperature interval of inversion effect (°C)	Hysteresis (°C)
2	200	No inversion effect to be detected				
3	236	Very broad, minima not exactly measurable	50 − 80	n. det.	~130	
4	315	Very broad, but with clear two minima (I, II)	I. 118 ± 3 II. 131 ± 3	I. 139 ± 3 II. 144 ± 3	99	− 21
5	372	Broad, but very clear two minima (I, II)	I. 187,5 ± 2 II. 203.5 ± 2	206 ± 2	70	− 18.5
6	410	Middle-broad, only one minimum (sharp)	213.5 ± 1	219 ± 2	25	− 5.5
7	508	Very sharp minimum (peak)	243.5 ± 0.5	233 ± 1	13	+ 10.5
8	613	Very sharp minimum	243.5 ± 0.5	232 ± 1	13	+ 11.5
Sample from HEIMANN	1300[a]	Very sharp minimum	268 ± 2	248 ± 2	...	+ 20

Hysteresis = t_i (heating) − t_i (cooling).
[a] p = 1 atm.

2. the width of the inversion temperature interval;
3. the shape of the inversion peak, and
4. the hysteresis between t_i measured during heating and subsequent cooling.

Besides the t_i, above all the heating-cooling hysteresis shows a very suitable measure for the degree of disorder, because it is not influenced by any grain size effect. With decreasing degree of disorder the t_i is continuously retarded, in heating as well as in cooling, because the number of structural defects which stimulate the inversion and act as "germs" of the new phase has been decreased. In DTA curves of strongly disordered crystals the t_i appears during cooling at a higher temperature than during preceding heating, the result being this kind of promotion of the inversion by structural defects; in this case the heating-cooling hysteresis shows a negative value. The hysteresis becomes increasingly positive with the improving degree of order, and the degree of order becomes increasingly better with increasing synthesis temperature (Fig. 73) if all other conditions are kept constant. The relatively low temperatures of

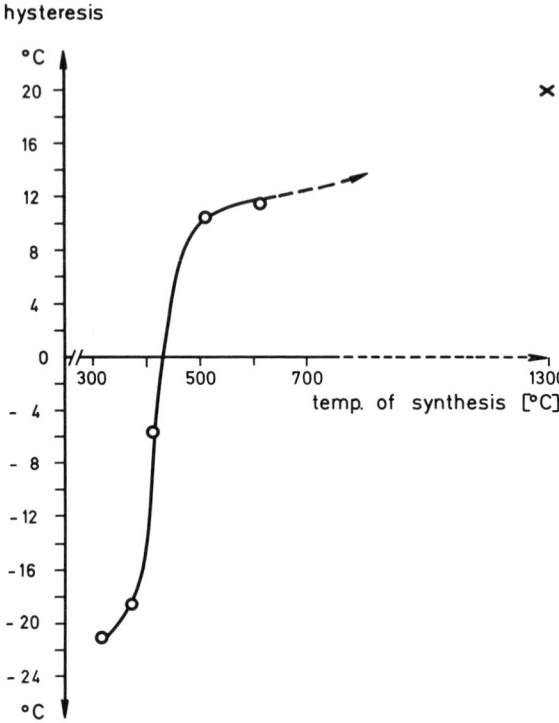

Fig. 73. The interdependence of hysteresis and synthesis temperature of cristobalites. One value from HEIMANN (X)

formation of these cristobalites (200–600° C) can be understood as a consequence of high SiO_2 supersaturations due to the rapid decay of Si_3N_4 under hydrous conditions. Substitution of NH_4^+ ions in cristobalite, which might behave like alkali ions (as described by FLÖRKE, 1955, 1961, 1962) in causing a lowering of synthesis temperatures as well as of inversion temperatures, has not been observed in additional infraredspectrometric investigations (for detailed discussion see SMYKATZ-KLOSS and SCHULTZ).

In 14 of the 17 samples mentioned above cristobalite was accompanied by chalcedony. Three samples of jasper (26a, 12a, 25) show the lowest t_i of all cristobalites having been investigated, and the t_i of the accompanying chalcedony (= microcrystalline quartz) is also very low (Fig. 74). It reflects the fact that both minerals are strongly disordered and so leads to the inference that the formation temperatures of both minerals should be very low, both having been formed close to the surface during diagenetic processes (compare with SMYKATZ-KLOSS, 1972a). The microcrystalline quartz samples of groups II and III (see

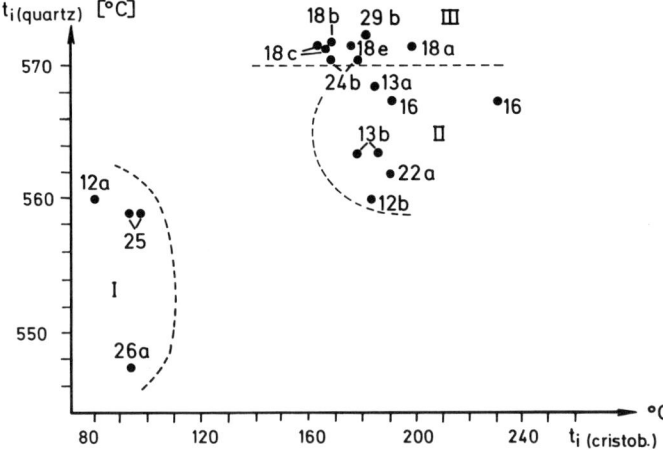

Fig. 74. t_i of naturally formed cristobalites versus t_i of accompanying microcrystalline quartz crystals

Fig. 74) contain disordered cristobalite, too, but the degree of disorder is not so great as in the case of cristobalite of group I. The broken line in Fig. 74 parallel to the abscissa showing the border between the t_i of quartz crystals formed magmatically or authigenically in sediments (see III-8) only enables a distinction of the chalcedony of groups II and III. The t_i of cristobalites of both groups does not differ.

The microcrystalline quartz samples 18 a–e show a t_i higher than 571.5° C, so pointing out that the agate sample 18 should have been formed by precipitation from hydrothermal solutions (compare with III-8 and -9). Since the cristobalite of this sample 18 show a certain degree of disorder (compare with Table 33), it could have been formed later, e.g. during diagenesis.

From shape, peak temperatures, width of temperature interval (see Table 34) and heating-cooling hysteresis of cristobalite inversions appearing in DTA curves of cristobalite-bearing rock samples, statements on the disorder character of these cristobalites can be made. The comparison of t_i of low-temperature cristobalites formed metastably in soils and sediments with t_i of accompanying quartz crystals can give some hints to the petrogenesis of these cristobalite-bearing rocks.

8. The Determination of Inversion Temperatures of Quartz Crystals as a Petrologic Tool

According to a proposal of FAUST (1948) in thermal analysis the temperature of the reversible structural transformation in second coordination ("displacive transformation", BUERGER, M.J.; ZEMANN) from trigonally crystallizing low- to hexagonal high-quartz lying at $\sim 573°$ C was frequently used for calibration of thermocouples, though FENNER and SOSMAN had assumed that these inversion temperature might vary for quartzes of different conditions of formation. Finally TUTTLE found out on investigating 30 quartz samples differential thermal analytically that the t_i of these quartz crystals were variable up to 1.9° C, and KEITH and TUTTLE stated on investigating 250 predominantly magmatically formed quartz crystals that the t_i of nearly 95% of these samples did not deviate more than $\pm 2°$ C from the "textbook-t_i" of 573° C (CORRENS, ZEMANN a. o.). The t_i of the remaining 5% could, however, appear down to 538° C. By comparing these values with the t_i of synthetic quartz crystals in which Si^{4+} had been partly substituted by Ge^{4+} or $Al^{3+} + Li^+$, and in which the margin of error in t_i-determination had sometimes been ± 5–15° C (!) KEITH and TUTTLE found out the following "rule": the inversion temperature of quartz crystals will be the lower, the higher the formation temperature of these crystals has been. As a reason for this "rule" they gave the argument that with increasing temperatures of formation the tolerance of the quartz structure for the incorporation of strange ions will also increase, so lowering the stability of the structure by increasing substitutions. KEITH and TUTTLE found numerous exceptions from this "rule" which they explained by "various surroundings of formation". The author's DTA investigations on more than 400 quartz samples from different localities (including the samples described in III-9), including many more samples from non-igneous rocks than KEITH and TUTTLE had available, indicate the very opposite: *the inversion temperature of quartz crystals from igneous rocks are higher than those from sedimentary rocks formed by processes of diagenesis or weathering.* The t_i of most metamorphic quartz crystals also lie above those from sediments, but the inversion behaviour of metamorphic quartz crystals is a little more complicated than that of igneous quartz crystals.

The diagram of Fig. 75 sums up the inversion temperature of nearly 300 quartz samples investigated under standard conditions, but with a nickel block sample holder and using K_2SO_4 or the mineral cryolite, Na_3AlF_6, as an internal standard. The exactness of measurement has been $\pm 0.3°$ C for t_i-values $> 570°$ C (for well-ordered quartz crystals with sharp inversion peaks) and ± 1–2° C for t_i-values $< 570°$ C (= for disordered quartz crystals from sediments, see below). More details about the method and about the different t_i-data of all samples can be found in SMYKATZ-KLOSS (1970–1972a). The broken line of Fig. 75 parallel to the ordinate subdivides the t_i of sedimentary formed quartz crystals and the t_i of igneous quartz crystals; it appears at $570.9 \pm 0.3°$ C ($\sim 571°$ C). As can be seen from Fig. 75, only the quartz crystals from

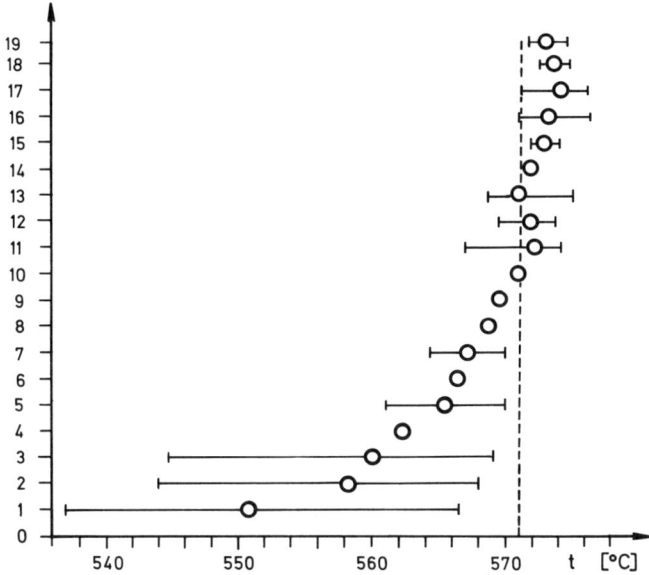

Fig. 75. Variation and average values of t_i of quartz crystals from igneous (16–19), metamorphic (12–15) and sedimentary rocks (1–11). (1) Slates with siliceous concretions ("Kieselgallen-Schiefer"), (2) diagenetic formations from cavities in volcanics (amygdales), (3) authigenic crystals from slate clays, (4) diagenetic matrix of quartzitic sandstones, (5) authigenic crystals from marls, (6) hornstones from dolomitic limestones, (7) authigenic crystals from sulfate rocks, (8) siliceous sinter, (9) authigenic crystals from limestones, (10) authigenic from coals, (11) sandstone quartzes, (12) crystals from regional metamorphic rocks, granulite facies, (13) amphibolite facies, (14) greenschist facies, (15) contactmetamorphic altered sandstone quartzes, (16) crystals from hydrothermal veins, (17) crystals from pegmatites, (18) crystals from plutonites, (19) crystals from volcanics

sandstones and those from regional metamorphic rocks show t_i on both sides of this "border line", but their average values lie on the "igneous" side. Quartz crystals from sandstones, originally products of weathering of igneous and metamorphic rocks, had come detritally into the sedimentation basins; so the similarity of sandstone quartzes with metamorphic and igneous quartz crystals can be explained. From Fig. 75 the t_i of sedimentary (S) and igneous (M) quartz crystals can be distinguished as follows:

$$t_{i(s)} < 571°\text{C} < t_{i(M)}.$$

Naturally occurring quartz crystals show variable t_i between nearly 500° and 578° C (III-9 and SMYKATZ-KLOSS, 1971). But the wide temperature range from 500° to 571° C is only occupied by the t_i of sedimentary formed quartz crystals and by the t_i of some metamorphic quartz crystals, together amounting to five to seven percent of all quartz crystals. And most of the quartz crystals occurring in natural rocks, e.g. more than 93% (= all crystals from igneous and most quartz crystals from metamorphic rocks), occupy only the small temperature range between 571° and 578° C. There may exist some exceptions from this scheme — but the author did not succeed in finding them.

The question is what can be responsible for this variation of inversion temperatures of natural quartz crystals? KEITH and TUTTLE mentioned three reasons: first the substitution of Si^{4+} by $Al^{3+} + (Li^+, Na^+)$, second inclusions of other material, and third "crystal physical factors"; according to KEITH and TUTTLE substitutions will be the predominant reason for t_i-lowering.

If a crystal is "homogeneous" (KEITH and TUTTLE) in wide ranges, that means without any notable unsteadiness, or "inhomogeneous", for instance by including numerous other particles, the t_i will differ: structural transformations begin at these inclusions. If this were correct, DTA curves of quartz crystals rich in inclusions ought to show broad inversion effects without a marked top, but only showing some small indistinct endothermic minima. This can be observed in DTA curves of some microcrystalline quartz samples (compare with III-9) or low-temperature cristobalites (compare with III-7). But the inversion effects of quartz crystals from igneous rocks and rich in inclusions did *not* differ from comparable crystals free of inclusions — the peaks being sharp, the temperatures above 571° C: so inclusions do not seem to be responsible for t_i-variations (compare with SMYKATZ-KLOSS, 1970).

According to investigations of BAMBAUER or BRUNNER, WONDRATSCHEK and LAVES or DENNEN a.o. the quartz structure only tolerates the incorporation of small contents of substituents (<0.3% of alkalies, Ca^{++}, Mg^{++}, Al^{3+}, H^+, traces of iron, manganese and titanium). The t_i of some sandstone quartzes rich in Li^+, however, appeared 2–3° C lower than those of samples lacking in Li (SMYKATZ-KLOSS, 1970), but the t_i-determination of 22 rock crystals from alpine clefts that were chemically analyzed by BAMBAUER, as well as the t_i-determination of some chemically analyzed microcrystalline quartzes (see III-9), had shown no relation between t_i and trace element contents of these quartz samples (compare with SMYKATZ-KLOSS, 1972a).

Investigations on tridymites and cristobalites made by FLÖRKE (1955) showed a t_i-dependence on the degree of disorder of these minerals (definition of disorder see III-7). And three samples of slates containing siliceous concretions that differed greatly in their widths at half height of X-ray interferences (KNOKE) showed with increasing disorder inversion temperatures of 566°, 549° resp. 537°±2° C (SMYKATZ-KLOSS, 1970). According to FLÖRKE (1962) "strong disorder generally points to low temperature of formation". Indeed the sedimentary quartz crystals which had formed during weathering or diagenesis show lower t_i than the investigated quartz crystals from pegmatites, granites, hydrothermal veins, rhyolites and other igneous rocks (see Fig. 75).

The degree of (dis-)order of a crystal is apparently a mirror reflecting all processes which have influenced the structural stability of this crystal, including processes of leaching and recrystallization, of pressure solution, micro-tectonics, weathering, diffusion and so on. Important will be the *"primary disorder": the more defects a forming crystal incorporates, the easier the crystal will be altered by subsequent processes of diagenesis or metamorphism.* If a crystal has only few primary defects, if the primary disorder is very slight, all subsequent processes will cause only very few alterations. All factors prevailing during weathering and diagenesis should be very favourable to produce structural disorder, but only in newly formed phases and not in detrital crystals from igneous or metamorphic rocks which have a very small primary disorder protecting them from alterations. The most important of these favourable factors are: low temperatures, pore solutions containing a large offer of substituents, (some of these solutions being chemically very aggressive), and geological time. From the view of kinetic and of chemical environment (= pore solutions and surrounding rocks), quartz crystals forming authigenically in sediments should involve more defects than crystals being formed during cooling from a melt.

Apart from the primary disorder being mainly responsible for the crystal physical properties (e.g. for the inversion temperature of quartz crystals), metamorphic processes including great variations of temperatures and pressures can produce *"secondary defects"* (MCDOUGALL), of which the dependence on temperature may be different from that of primary defects. Therefore metamorphism often causes very special conditions of disorder: at appointed temperatures several structural defects disappear by recrystallization ("critical temperature of recrystallization", BUERGER and WASHKEN), consequently the degree of disorder decreases, and at temperatures somewhat higher some new defects can be produced, the degree of disorder thus being increased. Besides temperature (and pressure) there may be several other resources of structural variations, factors influencing the structural stability of crystals like radioactive decay or changes in the composition of fluids or gases having been in equilibrium with a crystal. However, in the case of metamorphic rocks the complexibility of different defects in a crystal structure, effective and observable as "degree of disorder", show some obscurities which do not in all cases allow a clear petrogenetic interpretation of t_i-determinations of metamorphic quartz crystals. Consequently there will be only few

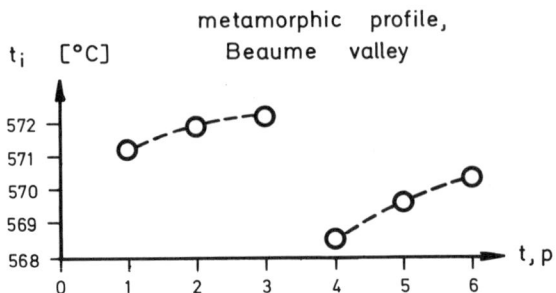

Fig. 76. t_i of quartz crystals from the metamorphic profile of the Beaume valley; grade of the metamorphism increasing from 1 to 6 (see TOBSCHALL, 1969, 1971)

examples of application of t_i-determination to the petrology of metamorphic rocks.

LAMEYRE et al. observed when investigating granites and surrounding mica schists from the Massif Central, France, that the quartz crystals of these granites showed inversion temperatures nearly 3° C lower than those of the surrounding mica schists or those of some other granites. Using the data given in the present study, the explanation of LAMEYRE'S observation must be that these granites and mica schists stand in some genetic relationship to each other. Both rocks may be products of the metamorphism of the same parent rock somewhere in the depths. Figure 76 contains the t_i of six quartz crystals from the metamorphic profile of the Beaume-valley (Cévennes, France), described by TOBSCHALL (1969, 1971), the rocks being characterized by an increasing grade of metamorphism from sample 1 to sample 6. According to TOBSCHALL the pre-metamorphic rock had been a greywacke. With increasing metamorphism the t_i of quartz crystals was also increasing, but only from sample 1 to 3 (Fig. 76).

Apparently the degree of disorder had been decreased from 1–3 by recrystallization (according to BUERGER and WASHKEN) with increasing temperatures of metamorphism (the pressure of all rocks of this profile had been nearly the same, compare with TOBSCHALL, 1969, 1971). The sudden decrease of t_i between sample 3 and 4 points to a greater change of metamorphic conditions producing new defects in the structure of quartz crystals. According to TOBSCHALL (1971) the rocks of the cordierite-amphibolite facies begin with sample 4 while the first three samples still belong to rocks of the greenschist facies. The further gradual increase of metamorphism (samples 5 and 6) seems to cause a new recrystallization of quartz crystals, consequently a decrease of the degree of disorder is reflected in a continuous increase of the inversion temperature.

The most extensive study of t_i of metamorphic quartz crystals has been made by KRESTEN (1971a and b), who corroborated the results of the author's investigations of igneous and sedimentary quartz crystals transferred to metamorphic rocks. By studying the inversion temperatures of quartz crystals from migmatites of the Västervik area of South Sweden, KRESTEN observed inversion temperatures continuously increasing with the increasing grade of metamorphism, showing the lowest t_i in greywackes that are only slightly metamorphic and the highest t_i in completely migmatized rocks. Mapping the inversion temperatures by means of "t_i-isotherms", KRESTEN obtained a good coincidence of inversion temperatures and petrographic findings.

The parallelism between the degree of disorder and inversion temperature and by this the indirect conclusion from t_i on the temperature of formation can be more easily drawn for quartz crystals from igneous and sedimentary than from metamorphic rocks. By optical investigation of thin sections of carbonate rocks containing idiomorphic quartz crystals it is often possible to conclude on the formation of these quartz crystals by searching for inclusions. Thus the comparison of optical observations (KORITNIG) with inversion temperatures showed that the well-known idiomorphic quartz crystals from Suttrop (Westfalia) had been formed from hydrothermal solutions, but the dark, so-called "stinking quartzes" from Dietlingen (Baden), which seemed very similar to the first, had been formed diagenetically (see SMYKATZ-KLOSS, 1969). In DTA curves of some siliceous sandstones (see III-9) the t_i of the diagenetically formed matrix between the sand grains is reflected in a distinct shoulder at temperatures 10–50° C lower than the t_i of the quartz grains of these sandstones. So samples from profiles of contact areas around basalts, going from red, uninfluenced sandstone to the faded and partly molten sandstone clods already included in basalt show in their DTA curves these shoulders. They represent the strongly disordered part of the quartz crystals which has been formed diagenetically, and shift to higher temperatures with decreasing distance to the basaltic contact until they disappear completely, immediately on contact. Apparently the "primary defects" of this sandstone matrix consisting of strongly disordered quartz had been recrystallized by the influence of the high temperatures of the basaltic melt which had reached nearly 1100° C at the contact to the surrounding sandstone (compare with SMYKATZ-KLOSS, 1971).

Clear, idiomorphic quartz crystals having formed authigenically in soils showed a very broad, not very intensive inversion deflection with a peak around 500° C (VAN DER PLAS). Several endothermic minima in DTA curves of quartz crystals from brown coals could be explained by comparing optical investigations: well-rounded detrital crystals free of inclusions showed zones of hypidiomorphic or xenomorphic secondary quartz rich in inclusions of brown coal, and grown diagenetically including the detrital grains (BEISING). Eight DTA runs of samples with different portions of detrital and authigenic quartz showed that the "shoulders" appearing below 570° C were caused by the inversion of these authigenically formed disordered quartz from the growth zones. In tempering experiments (heating the samples up to 650° C 30–50 times), no recrystallization of these disordered secondary quartz could be observed, but only a continuous disappearing (after the 20th run) of the K_2SO_4-inversion peak used as an internal standard.

While the t_i of *synthetic quartz* crystals could only be determined with a large margin of error ($\pm$ 5–15° C, compare with KEITH and TUTTLE), GILLERY was able to point out a distinct dependence of the various t_i of synthetic $AlPO_4$-compounds, all isotype with related SiO_2-phases, on the synthesis temperature: he found increasing inversion temperatures with increasing synthesis temperatures, as has also been described for cristobalites in this study.

All the examples mentioned demonstrate that the determination of inversion temperatures of quartz crystals can contribute to solving problems of sedimentary petrography or petrology. So this method can be used for the distinction of authigenic and detrital quartz crystals from sediments which might not be simple if no inclusions are to be found, and if the crystals are xenomorphic. It can also be used to determine the derivation of detrital quartz crystals, which in special cases may be helpful in prospecting for minerals, and generally for the characterization of quartz-bearing rocks. For detailed discussion of these and more examples see SMYKATZ-KLOSS, (1970–1972b), BEISING or KRESTEN, (1971a, b).

9. The High-Low Inversion Behaviour of Microcrystalline Quartz Crystals

The temperatures of structural transformations of (quartz) crystals are dependent on the degree of disorder, as has been discussed previously (III-7, 8). According to FLÖRKE (1961) micro- and cryptocrystalline particles, being finer than a distinct grain size, should be influenced in their structural (dis-)order by the large relation particle surface/volume, because the surface of a crystal is always disturbed. Provided that the crystal to be examined is a cube with a disturbed surface layer thickness of 10 Å, after FLÖRKE 6 volume-% of this cubic cristal will be disordered if the edge length of this crystal is 0.1 μ. In the case of an edge length of 0.01 μ, the disordered portion caused by the disturbed surface layer will be 50 volume-% of the particle. Therefore a distinct t_i-dependence on the size of quartz crystallites must be expected in the case of extremely finegrained samples. Three examples given by FLÖRKE (1961) point this out. However, data from KEITH and TUTTLE or WARNE as well as the author's investigations reported below show that this t_i-dependence on the size of crystallites is only valid for extremely fine-grained cryptocrystalline quartz particles, while numerous quartz crystals coarser than 0.05 µ ⌀ really invert at "normal" temperatures ($>570°$ C). This demonstrates that the portion of disorder coming from the disturbed surface layer can be neglected at least for these crystals showing t_i $>570°$ C. Thus in the case of micro- and cryptocrystalline quartz samples coarser than 0.05 µ ⌀ (in the following shortly named as "microcrystalline"), the main influence on t_i seems be exercised by *that* portion of structural disorder which has been caused by conditions of formation. Therefore the t_i of microcrystalline quartz samples (that means: of crystals with particle sizes between 0.05 µ and 200 µ ⌀ !) can apparently be expected as a measure for the conditions of formation, in the same way as has been discussed for macrocrystalline quartz crystals ($= >200$ µ ⌀, compare with III-8).

The material investigated in this study has been "chalcedony in a further sense" according to RAMDOHR and STRUNZ: microcrystalline quartz occurring in compact, homogeneously appearing aggregates which consist of very fine fibres ("chalcedony in a strong sense": the crystallographic c-axis of these microcrystalline quartz fibres are perpendicular to the length axis of the fibres) or of extremely small (~ 0.05–0.1 µ ⌀) isometric grains (in the case of jasper, chrysoprase, flints and hornstones). Some so-called "jaspers" were in fact rocks formed by fritting or silification (according to RAMDOHR and STRUNZ: banded jasper, ball jasper or basaltic jasper).

The most striking characteristic in DTA curves of microcrystalline quartz samples is the *shape* of the inversion deflection, which is generally very broad, has no marked top and is only of minimum intensity (ΔT: 0.3–0.8° C compared with 0.8–2.0° C in DTA curves of most macrocrystalline quartz samples). Frequently this inversion deflection of microcrystalline quartz shows *several* small peaks, none of which is very distinct (see Figs. 77 and 78). Even with a DTA apparatus of very high sensitivity, it often happens that no inversion can be recognized. BUURMAN and VAN DER PLAS (1971), for instance, only observed a clear inversion in the DTA curves of 3 hornstones and flints out of 24 samples, and Hoss described the inversion deflections of quartz crystals from siliceous shales as being very weak or not detectable. SOSMAN only observed the inversion of some chalcedonies from Mexico after he had raised the sensitivity of the apparatus as much as possible—the same samples had been described by FENNER as showing no inversion effect in

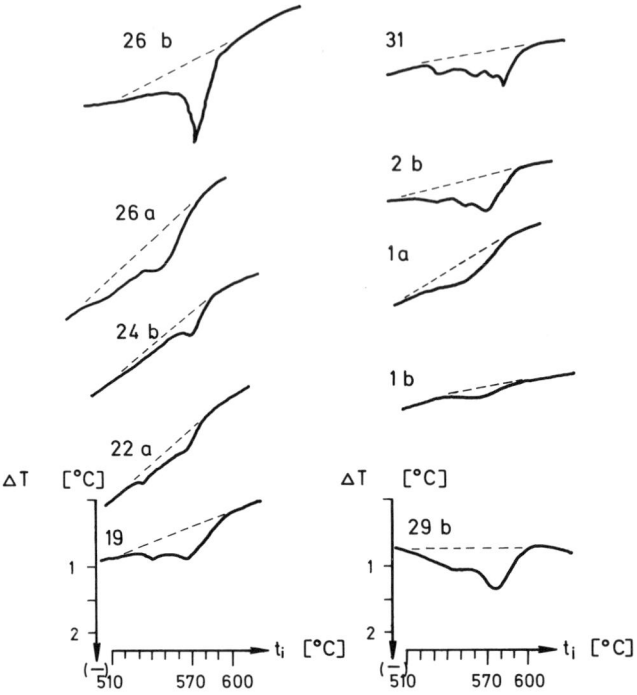

Fig. 77. DTA curves of some microcrystalline quartz samples (first heating, endothermic deflections. For comparison: the curve of one macrocrystalline quartz sample (26b); K_2SO_4 not having been drawn. For numbers of samples see Table 35

DTA. More DTA curves of such strongly disordered chalcedonies showing only very indistinct inversion effects are to be found in publications of FLÖRKE (1961); BETTERMANN; MOORE and ROSE; and KNELLER et al. In contrast to the broad shape of the t_i of microcrystalline quartz crystals the DTA curves of *coarser* quartz crystals ($>20\,\mu\,\varnothing$!) generally show only *one* inversion peak that is mostly very sharp and 3–5 times more intensive than the effect of microcrystalline samples (see Fig. 77, sample 26 b). For explanation of these broad inversion effects of most microcrystalline crystals, see III-7. Apart from the conditions of analysis mentioned in chapter III-8 in DTA investigations of chalcedonies and jaspers, the following must be taken into consideration: a comparable packing density of sample and inert material, an exact adjustment of the sample holder in the cylindric furnace, and an admixture of a smaller amount of internal standard material, K_2SO_4, than in the case of macrocrystalline samples. For a detailed description of the analytic conditions,

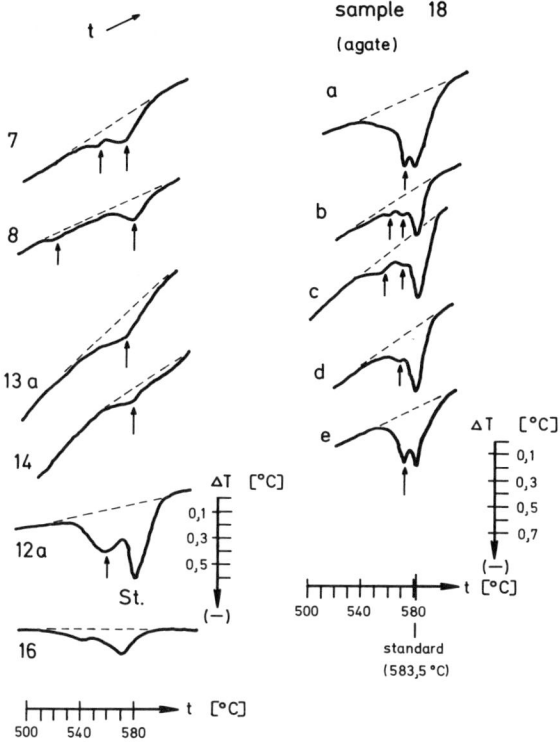

Fig. 78. DTA curves of some microcrystalline quartz samples (first heating). Curves 12a and 18a–e with K_2SO_4-inversion peak (St. in 12a)

see SMYKATZ-KLOSS (1972a). In order to recognize even very small inversion minima in the DTA curves each sample has been heated without the internal standard material. From these "blind runs" it became evident that the K_2SO_4 inversion peak has no influence on the t_i of microcrystalline quartz crystals. The peak temperatures listed in Table 35 have been measured indirectly from the peak temperature differences between the t_i of K_2SO_4 (= 583.5° C on heating) and that of quartz. By means of this indirect measurement and with the aid of a calibration curve peak temperature versus mV obtained by heating some calibration materials (K_2SO_4, Na_3AlF_6 and others, compare with Table 2), the exactness of temperature determination was ±0.3° C for less disordered crystals (= t_i > 570° C) resp. ±2–3° C for greatly disordered samples which show no distinct inversion peak but a broad, not very intensive effect. FLÖRKE (1961), too, has referred to the difficulties in exactly determining the peak temperatures of microcrystalline quartz crystals.

Table 35. The t_i of microcrystalline quartz crystals (ΔT in all cases between 0.3 and 1°C)

Sample	Locality	Appearance	t_i (°C)	
1a	Agate from Klingenmünster, Germany	Dark red		559 ±2
1b	Agate from Klingenmünster, Germany	Light red core of 1a		566.3 ±1
2a	Agate from Klingenmünster, Germany	Colourless	529 ±1	574.1 ±1
2b	Agate from Klingenmünster, Germany	Milky white rim of 2a	527 ±2; 553.5 ±2	571.9 ±1
3	Agate from Mt. Kossakov/CSSR	Green, transparent	535.5 ±2	557.5 ±1
4a	Agate from Mt. Kossakov/CSSR	Light grey	544.5 ±2	573.0 ±0.3
4b	Agate from Mt. Kossakov/CSSR	Dark grey	558.5 – 560.5 ±3	
6	Flint, loc. unknown	Dark blueish grey	553 ±3	563 ±3
7	Hornstone, Zell/Baden, Germany	Brown-grey	523.5 ±3	571.6 ±1
8	Chrysoprase, Kosemitz/Silesia	Green	518.5 ±2	578.7 ±0.3
10-1	Agate, Assif Imini, Morocco	Grey, core of the agate		572.8 ±1
10-2	Agate, Assif Imini, Morocco	Milky-white bands		572.0 ±1
10-3	Agate, Assif Imini, Morocco	Blue-grey	553 ±2	573.7 ±1
10-4	Agate Assif Imini, Morocco	Green, transparent	551 ±2	572.0 ±1
10-5	Agate, Assif Imini, Morocco	Blue-grey, rim of agate	557.7 ±2	569.0 ±1
12a	Jasper, Schweikhausen/Germany	Violet core of jasper		560.2 ±2
12b	Jasper, Schweikhausen, Germany	Flesh-coloured rim of jasper		560.3 ±2

13a	Ball jasper, Niederweiler/Germany	Beige		568.5 ± 2
13b	Ball jasper, Niederweiler/Germany	Red core of 13a		563.5 ± 2
14	Flint, Mendon n. Paris/France	Grey		573.3 ± 2
16	Agate, loc. unknown	Blueish-grey	539.5 ± 2	567.5 ± 2
18a	Agate, loc. unknown	Dense, grey rim		571.5 ± 0.3
18b	Agate, loc. unknown	Light red	558 ± 2	571.8 ± 2
18c	Agate, loc. unknown	Red and white bands	563 ± 3	571.5 ± 0.7
18d	Agate, loc. unknown	Dark brown		570.7 ± 1
18e	Agate, loc. unknown	Dense, grey core		571.6 ± 0.5
19	Agate, loc. unknown	Light grey	542 ± 2	568.5 ± 2
22a	Banded jasper, loc. unknown	Flesh-coloured	533 ± 2	562.0 ± 2
24b	Jasper, Eggenstein/Baden, Germany	Light brown		570.5 ± 2
25	Basalt jasper, Eichstetten, Germany	Green		559.0 ± 3
26a	Jasper, Tiefenstein/Nahe, Germany	Pink and white rim		547.3 ± 2
26b	Jasper, Tiefenstein/Nahe, Germany	Quartz from the core of 26a		574.1 ± 0.3
27	Hornstone from Piemont/Italy	Brownish black		573.5 ± 2
29a	Ball jasper, loc. unknown	Brown		571.5 ± 3
29b	Ball jasper, loc. unknown	Light brown core of 29a		572.5 ± 2
31	Chalcedony from brown coal, Frimmersdorf/Germany (see BEISING)	Light blue, grape-like	530 ± 2; 557 ± 1	573.3 ± 0.5

From 51 investigated samples in the DTA curves of 15 samples no inversion could be detected. In Table 35 the data of the remaining 36 samples have been listed, all having been heated according to standard conditions of analysis with the exception of those factors mentioned above. The DTA curves of 16 samples show two endothermic "peaks" due to the inversion of quartz, and two curves even contain three endothermic peaks caused by the inversion of microcrystalline quartz with a different degree of disorder.

Seven samples listed in Table 35 were chemically analyzed (Mrs. U.-LENNARTZ, Institute of Petrography, Universität Karlsruhe).

The amounts determined of Fe_2O_3 (0.1–1.9 weight-%) can indeed explain the brownish or reddish colours of some jaspers or agates (caused by finely-dispersed hematite but not by the incorporation of Fe^{3+} into the quartz structure!), but an interdependence between the amounts determined of MgO, Fe_2O_3, Na_2O, K_2O, Li_2O and the inversion temperature could not be found (see SMYKATZ-KLOSS, 1972a). It is, of course, hardly possible to distinguish between ions that are truly incorporated into the quartz structure and submicroscopical inclusions, even in the case of macrocrystalline quartz crystals (BAMBAUER).

The observation of FLÖRKE (1955), who found the inversion deflection of disordered tridymites and cristobalites very broad, "smeared over" a temperature interval of 30–50° C and not very intensive, can be corroborated by the author's DTA curves of many microcrystalline quartz crystals. Most of the samples investigated contain a portion of quartz with various degrees of disorder besides a portion of well-ordered quartz. But for crystals coarser than 0.05 μ ∅ there is no connection to be seen between t_i and the size of crystallites, as should be expected according to FLÖRKE (1961). This can be seen in some samples with particle sizes between 0.05 and 0.1 μ ∅ (determined by means of electron optics), which showed t_i-values higher than 570° C, for instance the samples 7 and 27 (hornstones), 8 (chrysoprase), 14 (flint), 24b (jasper) and 18 b–d (agates, compare with Table 35). Besides this main inversion peak > 570° C, some of the samples show small shoulders at lower temperatures representing the inversion of their disordered portions (see Figs. 77, 78).

Since considerable disorder of crystals generally indicates low temperatures of formation (FLÖRKE, 1962), it is evident that the low t_i of most microcrystalline quartz crystals can be correlated with low temperatures of formation. Of course, it has to be considered that disordered microcrystalline quartz crystals may also be formed by very quick cooling from melts. This case should be recognized by means of other petrographic criteria. However, microcrystalline quartz samples with t_i higher than 570° C, only partly disordered, should be formed or altered at higher temperatures by precipitation from hydrothermal solutions. The discussion of some examples may demonstrate the application of the method to petrographic work.

The formation of agates in amygdules of eruptive rocks usually takes a long time (FISCHER), the supplies of postmagmatic solutions often being interrupted by processes of resolution and diagenetic transformations. This can frequently be seen in small fissures and cracks serving as "channels of diffusion" (FISCHER), now filled up with secondary chalcedony. The occurrence of less disordered and very much disordered microcrystalline quartz in the same agate layer can be explained by such diagenetic processes.

These processes of diagenesis can be studied in a more detailed way by regarding the inversion behaviour of the various kinds of microcrystalline quartz crystals that have been formed diagenetically in eruptive rocks or sediments. Two different processes of diagenetic quartz formation must be considered. The first is only a simple precipitation of silica from SiO_2-bearing solutions circulating through porous near-surface rocks and so filling up caverns, clefts, fissures, amygdules, pores and so on (the process being called „Einkieselung" in German, see CORRENS), and the second process is a real silification including chemical reactions like resolution and replacement phenomena („Verkieselung"). The formation of agates that are precipitated as silica gels in already existing amygdules and cavities and then aged belongs to the first process, the formation of hornstones or flints having been formed after resolving some carbonateous matter belongs to the second one. A detailed investigation of 37 samples from 14 localities in the Black Forest, South Germany, brought some surprising results demonstrating that the method of t_i-determination will enable the distinction of both kinds of diagenetically formed microcrystalline quartz (compare with SMYKATZ-KLOSS, 1972b). The samples had been selected in such a way, that the portion of diagenetically formed quartz was as large as possible. For comparison some quartz crystals of igneous rocks were investigated, too (samples 1–5, Table 36). The data of Table 36 were obtained under standard conditions of analysis using K_2SO_4 for internal standard, the exactness of temperature determination being $\pm 1°$ C for $t_{i(d)}$ ($= t_i$ of the diagenetically formed portion of the quartz), resp. $\pm 0.3°$ C for $t_{i(m)}$ ($= t_i$ of the magmatically formed portion of the sample of the parent rock).

The DTA curves of these samples (Figs. 79–81) show very clearly that the diagenetically formed portion of the samples (marked with arrows) are disordered compared with the well-ordered quartz crystals of the parent rocks. Some curves even contain three quartz inversion peaks besides that intensive peak appearing at the highest temperature and due to the inversion of the K_2SO_4 standard material. Figure 81 contains the

Table 36. t_i of microcrystalline quartz crystals from fissures and veins and t_i of quartz crystals from some parent rocks (°C); all samples from localities in the Black Forest, Germany

Sample	Locality	$t_i(d)$	$t_i(m)$
2	Pegmatitic quartz from Friesenberg-Granite near Baden-Baden	—	573.0
3	Quartz from hydrothermal vein, Thimos	—	571.0
9	Silicified limestone, Badenweiler	543.5	573.2
12	Carneol-dolomite, Baiersbronn	563.7	572.3
1	Sandstone with diagenetic quartz matrix, Hornisgrinde	553	572.3
1a	Quartz from the rim of a fissure in 1	544	571.2
1b	Grown on 1 a	546.5	575.9
1c	from the center of the fissure	559	573.2
4	Sandstone with diagenetic quartz matrix, Würzbach	557	571.8
4a	Fissure in 4	—	571.3
5	Bunter sandstone, Grunbach	—	571.7
5a	Fissure in 5 + impurity of 5	554	571.3
6_{Ia}	Oldest vein quartz (edge of the vein), Neubulach	540; 559	571.5
6_{Ib}	Younger vein quartz; pure, grown on I a	—	573.3
6_{II}	Hypidiomorphic vein quartz, Neubulach	—	573.7
6_{III}	Hypidiomorphic vein quartz, Neubulach	541.4	575.2
6_{IV}	Hypidiomorphic vein quartz, Neubulach	539.5	571.0
7a	Xenomorphic quartz, edge from a vein, Enzklösterle	551.1	574.9
7b	Hypidiomorphic quartz, center of the vein, Enzklösterle	552.9	571.3
8_{Ia+b}	Xeno- to hypidiomorphic quartz from the edge of the ore vein „Friedrich-Christian"	544.7; 564.0	573.8
8_{Ic}	Center of the same vein	558.7	573.3
8_{IIa}	Second sample of this ore vein, from the edge	544.7	572.3
8_{IIb}	Hypidiomorphic, grown on a	561.0	574.5
8_{IIc}	Hypidiomorphic, center of the vein	554.2	573.3
11a	Xenomorphic, from the edge of a vein, St. Blasien	—	573.7
11b	Grown on a, pseudomorphic after baryte	—	573.5
11c	Grown on b, xenomorphic	551.8	572.0
11d	Idiomorphic crystals from the center of the vein	—	573.8
13a	Edge of a fluorite + chalcopyrite bearing ore vein, Brandenberg	554.9	572.7
13b	Center of this vein, hypidiomorphic, clear crystals	538.1	572.6
$10a_1$	Ore vein „Gottesehre" n. Urberg, a_1: clear, xenom., center	532.5; 562.4	
$10b_1$	With fluorite, calcite, chalcopyrite, b_1: milky, xenomorphic	539.7; 565.5	575.5
$10c_1$	with Galena, c_1: long-stalked	545.0; 563.7	574.7
$10d_1$	d_1: edge of vein, coarse-grained, xenomorphic	557.3	
$10a_2$	a_2: center, like a_1	569.8	
$10b_2$	b_2: milky + red, xenomorphic	569.8	
$10c_2$	c_2: edge, microcrystalline	553.0; 559.2	572.0

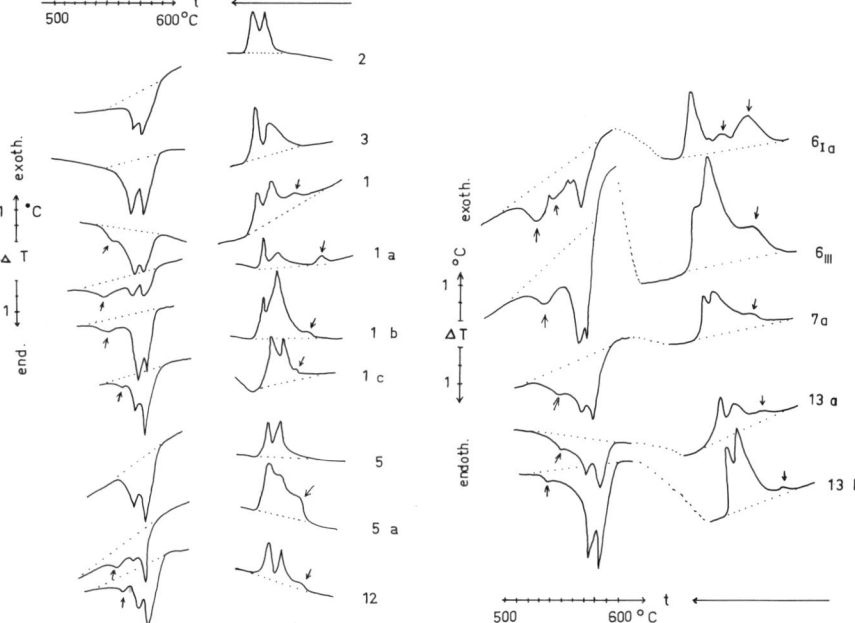

Figs. 79 (left) and 80 (right). DTA curves of quartz crystals from the Black Forest, Germany, showing the comparison of the inversion behaviour of igneous quartz and of microcrystalline quartz crystals formed diagenetically (marked by arrows). First heating (left side of each figure) and subsequent cooling (right side of each figure)

DTA curves of the different quartz generations of a hydrothermal fluorite-carbonate-baryte-ore minerals-quartz vein. This vein is five cm thick and its different varicoloured bands are arranged symmetrically from the middle of the vein (samples a) to the margins. It is surprising that the lowest t_i (532.5±1° C, see Table 36) belonged to the microcrystalline quartz a_1 occurring in the middle of the vein.

The petrogenetic interpretation of the t_i-data of the quartz crystals from cavities and veins has to be done in the following way. Most of these quartz samples were formed by precipitation from hydrothermal solutions, and naturally even the samples of the bunter sandstones (samples 1, 4, 5) and the pegmatite (sample 2) are of magmatogene origin; proof of this is that their main inversion peak appears at temperatures $>571°$ C ($= t_{i(m)}$ in Table 36). Of greater petrological interest than the conclusion on the formation will be that conclusion on all the diagenetic processes which can be drawn from the disordered parts of the quartz

156 Part III. Special Application of Differential Thermal Analysis

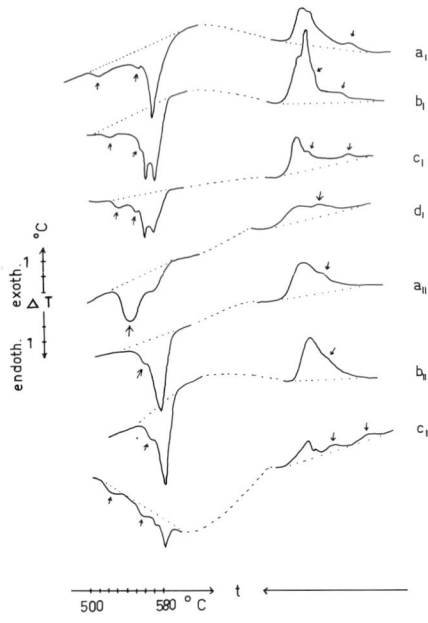

Fig. 81. DTA curves of the varicoloured layers (bands) of a fluorite-carbonate-baryte-quartz vein (sample 10, see Table 36); samples arranged symmetrically from the middle of the vain (a_1, a_2) to both margins (d_1, c_2). Left: first heating, right: subsequent cooling. Inversions due to diagenetically formed quartz marked by arrows. With K_2SO_4-inversion (peak at highest temperature)

samples reflected in the DTA curves in the inversion peaks that appear at lower temperatures and are not very intensive (with the exception of samples 6_{Ia}, 6_{III}, $10\,b_2$, $10\,d_1$). So at any time the cavities and pores of the sandstones could be filled up by secondary quartz forming the microcrystalline matrix between the macrocrystalline sandstone grains. For sample 1 the moment of this filling up can be read from $t_{i(d)}$-data lying between $1b$ and $1c$ (see Table 36), so being late-diagenetic: the pores of this bunter sample (represented by the $t_{i(d)}$-value of sample 1) were obviously filled up when the larger cavities (clefts, fissures) had already been occupied almost completely by secondary quartz (compare the $t_{i(d)}$-values of $1a$–$1c$). Another bunter sample (5) evidently had a lot of large cavities sufficient as capacity space for all secondary quartz, because the pores of this sandstone sample remained empty (which can be seen by the absence of a $t_{i(d)}$-value in sample 5).

This trend of evolution established in the sandstone sample 1 can also be recognized in the diagenetically formed quartz portions of some ore veins (Fig. 82):

The High-Low Inversion Behaviour of Microcrystalline Quartz Crystals 157

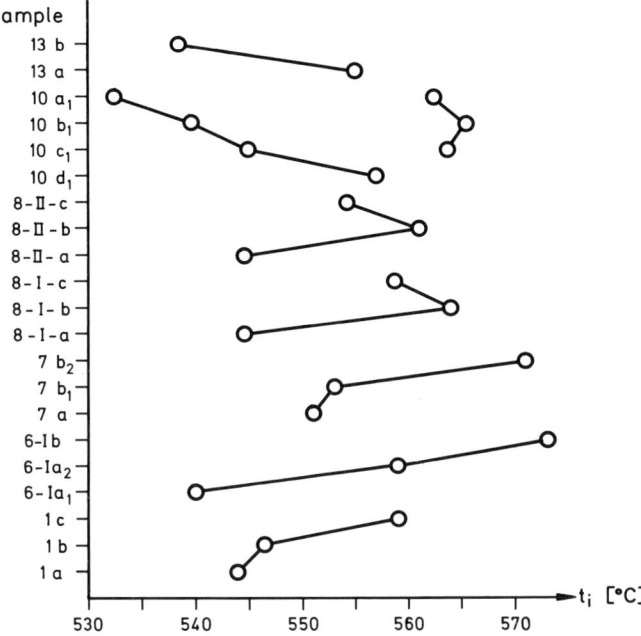

Fig. 82. t_i-differences of the various quartz generations formed diagenetically in veins and porous sediments

Generally the diagenetic quartz from the margins of a vein shows the lowest t_i, that from the middle of the vein the highest one. This is valid for all samples studied, with the exception of the samples 10 and 13 in which the relations were inverse, showing the lowest t_i not at the margins but in the middle of the vein. This apparent contradiction can be explained by optical investigations: with the exception of samples 10 and 13 all others were formed by filling up of the cavities, fissures and pores, so belonging to the first type mentioned above and characterized as *"normal"* quartz veins („Einkieselung"). These show the oldest diagenetic quartz formations with the lowest t_i, and these oldest diagenetic quartz had obviously been precipitated at the margins of existing cavities. Samples 10 and 13 still have a divergent mineral association primarily consisting of fluorite, baryte, calcite, dolomite, chalcopyrite, galena and only very few quartz. The carbonates and above all the baryte had then partly been replaced by secondary microcrystalline quartz, this silicification beginning in the *middle* of the veins where the only small cavities existed allowing a circulation of diagenetic pore solutions. From the middle of these hydrothermally formed veins (10 and 13), the replacement of primary vein minerals proceeded towards the vein margins, so being the second type of diagenetic quartz formation mentioned above (silicification, „Verkieselung") and forming a kind of *inverse* vein. This can be concluded from several pseudomorphs of microcrystalline quartz after baryte, from inclusions and relics. That means: in this case of silicification, the oldest diagenetic quartz generation will also show the lowest t_i, but this oldest diagenetic quartz is not to be found at the margins of the veins (as is the case in the cavity fillings), but in the middle of the veins.

The question of the *reason* for the t_i-differences of the different diagenetic generations from the same vein or cavity still cannot be answered completely. As the grain size for microcrystalline vein quartz samples does not differ greatly, this *cannot* be the reason, as discussed previously. The formation temperatures may have been somewhat different because the first pore solutions possibly deriving from postmagmatic solutions delivering the oldest diagenetic quartz generation cooled on meeting the host rocks, while a second similar solution following immediately after the first one did not cool as much as the first. Or perhaps the microcrystalline quartz crystals which were formed during the latest stages of diagenesis had more time to build up a better-ordered quartz structure than could the crystals of the first diagenetic stage. In any case, the recognition of that microcrystalline diagenetic quartz showing the lowest t_i occurring either at the margins or in the middle of a vein will be a suitable tool of petrogenetic interpretation, making possible the distinction between both types of diagenetic formation of microcrystalline quartz crystals.

References

ABELEDO, M. DE, ANGELELLI, V., BENYACAS, M. DE, GORDILLO, C.: Sanjuanite, a new hydrated basic sulfate-phosphate of aluminum. Am. Mineralogist **53**, 1–8 (1968).
AGAFONOFF, V.: Etude minéralogique du sol. Trans. 3rd. Internat. Congr. Soil Sci. **3**, 74–78 (1935).
AGIORGITIS, G.: Über differential-thermoanalytische und infrarotspektroskopische Untersuchung von Mangan-Mineralen. Tschermaks Mineral. Petrogr. Mitt. **13**, 273–283 (1969).
ALBEE, A. L.: Relationships between the mineral association, chemical composition and physical properties of the chlorite series. Am. Mineralogist **47**, 851–870 (1962).
ARENS, P. L.: A study of the differential thermal analysis of clays and clay minerals. Diss. Univ. Wageningen (Niederlande); Photodruck Excelsior-Offset, s'Gravenhage 1951.
ARISTARAIN, L. F., HURLBUT, C. S., JR.: Ameghinite, a new borate from Argentina. Am. Mineralogist **52**, 935–945 (1967a).
ARISTARAIN, L. F., HURLBUT, C. S., JR.: Macallisterite, $2MgO \cdot 6B_2O_3 \cdot 15H_2O$, from Salta, Argentina. Am. Mineralogist **52**, 1776–1784 (1967b).
ARISTARAIN, L. F., HURLBUT, C. S., JR.: Teruggite, a new mineral from Jujuy, Argentina. Am. Mineralogist **53**, 1815–1827 (1968).
ASENSIO, J., SABATIER, G.: Analyse thermique différentielle de quelques minéraux sulfurés et arséniés de fer, nickel et cobalt. Bull. Soc. Franc. Minéral. Cryst. **81**, 12–15 (1958).
BALL, D. F.: Interstratified illitic clay in Ordovician ash from Conway, N. Wales. Clay Minerals **7**, 363–366 (1968).
BAMBAUER, H. U.: Spurenelementgehalte und γ-Farbzentren in Quarzen aus Zerrklüften der Schweizer Alpen. Schweiz. Mineral. Petrog. Mitt. **41**, 335–369 (1961).
BARSHAD, I.: Vermiculite and its relation to biotite as revealed by base exchange reactions, X-ray analysis, d.t. curves, and water content. Am. Mineralogist **33**, 655–678 (1948).
BARTON, P. B., JR.: Thermochemical study of the system Fe-As-S. Geochim. Cosmochim. Acta **33**, 841–857 (1969).
BASSETT, W. A.: Copper vermiculites from N. Rhodesia. Am. Mineralogist **43**, 1112–1133 (1958).
BASTA, E. Z., KADER, Z. A.: The mineralogy of Egyptian serpentinites and talc-carbonates. Mineral. Mag. **37**, 394–408 (1969).
BAYLISS, P.: Effect of particle size on d.t.a. Nature **201**, 1019 (1964).
BECK, C. W.: An amplifier for differential thermal analysis. Am. Mineralogist **35**, 508–524 (1946).
BECK, C. W.: D.t.a. curves of carbonate minerals. Am. Mineralogist **35**, 985–1013 (1950).

BEISING, R.: Die Minerale der niederrheinischen Braunkohle und ihr Verhalten bei der Verbrennung in Kraftwerken. Neues Jahrb. Mineral. Abh. **117**, 1, 96–115 (1972).

BERG, L. G.: Einführung in die Thermographie. Moskau: Verl. Akad. Wiss. d. UdSSR 1961.

BERG, L. G.: Simple Salts; chapter 11, In: MACKENZIE, R. C. (Ed.): Differential Thermal Analysis, Vol. 1, pp. 343–361 (1970).

BERG, L. G., RASSONSKAJA, I. S.: Thermo-Analyse als Schnellverfahren. Dokl. Akad. Nauk. USSR **73**, 113–125 (1950).

BERKELHAMER, L. H., SPEIL, S.: Differential Thermal Analysis. Mine Quarry Eng. **10**, 221–225, 273–279 (1945).

BETTERMANN, P.: Untersuchung zur Mineralzusammensetzung und Kristallinität potentiell alkaliaktiver Feuersteine aus Schleswig-Holstein. Neues Jahrb. Mineral. Abh. **120**, 1, 51–68 (1973).

BIDLÓ, G.: Mineralogical analysis of Dunaujvaros Pleistocene drilled samples. Periodica Polytechn. Hung. Civil Eng. **15**, 1, 1–11 (1971).

BLUM, S. L., PALADINO, A. E., RUBIN, L. G.: D.t.a. techniques for determining Curie points. Am. Ceram. Soc. Bull. **36**, 175–176 (1957).

BOERSMA, S. L.: A theory of d.t.a. and new methods of measurements and interpretation. J. Am. Ceram. Soc. **38**, 281–284 (1955).

BOLLIN, E. M.: Chalcogenides. Chapter 7 of "Differential Thermal Analysis" by R. C. MACKENZIE, Vol. 1, pp. 193–234 (1970).

BORST, R. L., KATZ, J. L.: Epigenetic chlorite crystals in faulted lower Devonian mudstones, Kingston, N.Y. Am. Mineralogist **55**, 1359–1373 (1970).

BOSS, B. D.: D.t.a. of biotitic vermiculite to determine vermiculite content. Am. Mineralogist **52**, 293–298 (1967).

BRADLEY, W. F., BURST, J. F., GRAF, D. L.: Crystal chemistry and d.t. effects of dolomite. Am. Mineralogist **38**, 207–217 (1953).

BRAITSCH, O.: Über die natürlichen Faser- und Aggregationstypen beim SiO_2 und ihre Verwachsungsformen, Richtungsstatistik und Doppelbrechung. Heidelberger Beitr. Mineral. Petrogr. **5**, 331–372 (1957).

BRINDLEY, G. W.: The crystal structure of some chamosite minerals. Mineral. Mag. **29**, 502–522 (1951).

BRINDLEY, G. W.: Kaolin, Serpentin, and Kindred Minerals. In: BROWN, G. (Ed.): The X-ray identification and crystal structures of clay minerals, pp. 51–131, 2nd ed. London: Mineral. Soc. 1961.

BROUSSE, R., GASSE-FOURNIER, G., LEBOUTEILLER, F.: Cristaux de rozénite et de mélantérite dans la mine de diatomites de la Bade (Cantal). Bull. Soc. Franç. Minéral. Cryst. **89**, 348–352 (1966).

BROUSSE, R., GUÉRIN, H.: Découverte de pickéringite dans le Cantal. Bull. Soc. Franç. Minéral. Cryst. **88**, 704–705 (1965).

BROWN, G. (Ed.): The X-ray identification and crystal structures of clay minerals. London: Mineral. Soc. 1961.

BRUNNER, G. L., WONDRATSCHEK, H., LAVES, F.: Über die Ultrarotabsorption des Quarzes im 3 µ-Gebiet. Naturwissenschaften **24**, 664 (1959).

BUERGER, M. J.: Crystal structure analysis. New York: Wiley & Sons 1960.

BUERGER, M. J., BUERGER, N. W.: Low chalcocite and high-chalcocite. Am. Mineralogist **44**, 55–65 (1944).

BUERGER, M. J., WASHKEN, E.: Metamorphism of minerals. Am. Mineralogist **32**, 296–308 (1947).

BUERGER, N. W.: The chalcocite problem. Econ. Geol. **36**, 19–44 (1941).

BUURMAN, P., VAN DER PLAS, L.: The occurrence of halloysite and gibbsite in peneplain deposits of the Belgium Coudros. Geol. Mijnbouw **47**, 345–348 (1968).

BUURMAN, P., VAN DER PLAS, L.: The Genesis of Belgian and Dutch Flints and Cherts. Geol. Mijnbouw **50**, 9–28 (1971).

CABRI, L. J.: A new copper-iron sulfide. Econ. Geol. **62**, 910–925 (1967).

CAILLÈRE, S.: Contribution des mineraux des serpentines. Bull. Soc. Franç. Minéral. Cryst. **59**, 163–326 (1936).

CAILLÈRE, S.: Contribution et l'étude de l'hydromagnésite et des quelques autres hydrocarbonates magnésiens. Bull. Soc. Franc. Minéral. Cryst. **66**, 55–70 (1943 a).

CAILLÈRE, S.: Sur les caractères spécifiques d'un groupe d'hydrocarbonates magnésiens. Bull. Soc. Franc. Minéral. Cryst. **66**, 494–502 (1943 b).

CAILLÈRE, S., HÉNIN, S.: Verre et Silice Ind. **13**, 114–118 (1948).

CAILLÈRE, S., RODRIGUEZ, I.: Etude minéralogique de la Tinta formation argileùse de la province de Buenos Aires, Argentine. Bull. Soc. Franc. Minéral. Cryst. **90**, 246–251 (1967).

CAMPBELL, A. S., MITCHELL, B. D., BRACEWELL, J. M.: Effects of particle size, pH and organic matter on the thermal analysis of allophane. Clay Minerals **7**, 451–454 (1968).

CARPENTER, D. F.: Quantitative DTA. Proc. ICTA IV, Budapest 1974 (in press).

CARTHEW, A. R.: The quantitative estimation of kaolinite by d.t.a. Am. Mineralogist **40**, 107–117 (1955).

ČERNY, P., POVONDRA, P., STANEK, J.: Two cookeites from Czechoslovakia. Lithos **4**, 7–15 (1971).

CESBRON, F., FRITSCHE, J.: La mounanaite, nouveau vanadate de fer et de plomb hydrate. Bull. Soc. Franc. Minéral. Cryst. **92**, 196–202 (1969).

CHAO, G. Y.: Carletonite, $KNa_4Ca_4Si_8O_{18}(CO_3)_4(F,OH) \cdot H_2O$, a new mineral from Mt. St. Hilaire, Quebec. Am. Mineralogist **56**, 1855 (1971).

CHEN, P.-Y.: Mineralogical study of chromian thuringite from Shakatang-chi, Hualien. Acta Geol. Taiwanica **13**, 9–19 (1969).

CHEVALLIER, R., BOLFA, J., MATHIEU, S.: Titanomagnétites et ilménites ferromagnétiques. Bull. Soc. Franc. Minéral. Cryst. **78**, 307–346 (1955).

CHUKROV, F. V., BERKHIN, S. I., ERMILOVA, L. P., MOLEVA, V. A., RUDNITSKAYA, E. S.: Allophanes from some deposits of the USSR. Intern. Clay Conf. Stockholm 1963 Proc., pp. 19–28 (1963).

COHEN, L., KLEMENT, W., KENNEDY, G. C.: Journal of the physics and chemistry of solids. J. Phys. Chem. Solids **27**, 179–186 (1966).

COLE, W. F.: Interpretation of d.t.a. curves of mixed-layer minerals of illite and montmorillonite. Nature **175**, 384–385 (1955).

COLE, W. F., CROOK, D. N.: A note on the examination of pyrite in conventional d.t.a. equipment. Am. Mineralogist **51**, 499–502 (1966).

CORRENS, C. W.: Einführung in die Mineralogie. 2. Aufl. Berlin-Heidelberg-New York: Springer 1968.

CREMER, V.: Die Mischkristallbildung im System Chromit-Magnetit-Hercynit zwischen 1000 und 500° C. Neues Jahrb. Mineral. Abh. **111**, 184–205 (1969).

CUTHBERT, F. L.: Clay minerals in Lake Erie sediments. Am. Mineralogist **29**, 378–388 (1944).

CUTHBERT, F. L., ROWLAND, R. A.: D.t.a. of some carbonate minerals. Am. Mineralogist **32**, 111–116 (1947).

D'ANS, J., LAX, E.: Taschenbuch für Chemiker und Physiker. 3. Aufl. Berlin-Heidelberg-New York: Springer 1967.

DASGUPTA, D. R.: Thermal decomposition of dolomite and ankerite. Mineral. Mag. **36**, 138–141 (1967).

DEEG, E.: Grundlagen zur theoretischen Behandlung der DTA. Folgerungen für die Praxis. Ber. Deut. Keram. Ges. **33**, 321–329 (1956).

DENNEN, W. H.: Stoichiometric substitution in natural quartz. Geochim. Cosmochim. Acta **30**, 1235–1241 (1966).
DICHTL, H. J., JEGLITSCH, F.: Kombination der Hochtemperaturmikroskopie mit der DTA. Radex-Rundschau **1967**, H. 3/4, 716–722 (1967).
DIETRICH, V.: Ilvait, Ferroantigorit und Greenalith als Begleiter oxidisch-sulfidischer Vererzungen in den Oberhalbsteiner Serpentiniten. Schweiz. Mineral. Petrog. Mitt. **52**, 57–74 (1972).
DJURLE, S.: An X-ray study on the system Cu-S. Acta Chem. Scand. **12**, 1415–1426 (1958).
DUFF, E. J.: Magnetic interactions in the olivine structure. J. Chem. Soc. A, 2072–2074 (1968).
DUNNE, J. A., KERR, P. F.: An improved thermal head for d.t.a. of corrosive materials. Am. Mineralogist **45**, 881–883 (1960).
DUNNE, J. A., KERR, P. F.: DTA of Galena und Clausthalite. Am. Mineralogist **46**, 1–11 (1961).
DUVAL, C.: Inorganic Thermogravimetric Analysis, 2nd. Edition. Amsterdam: Elsevier Publ. Co. 1963.
EARLEY, J. W., MILNE, I. H., MCVEAGH, W. J.: Thermal, dehydration, and X-ray studies on montmorillonite. Am. Mineralogist **38**, 770–773 (1953).
EARLEY, J. W., OSTHAUS, B. B., MILNE, I. H.: Purification and properties of montmorillonite. Am. Mineralogist **38**, 707–724 (1953).
ECHLE, W.: Loughlinit und Analcim in neogenen Sedimenten Anatoliens. Contrib. Mineral. Petrol. **14**, 86–101 (1967).
ECKARDT, F. J.: Über Chlorite in Sedimenten. Geol. Jahrb. **75**, 437–474 (1958).
ECKARDT, F. J.: Über den Einfluß der Temperatur auf den kristallographischen Ordnungsgrad von Kaolinit. In: ROSENQUIST-GRAFF-PETERSEN: Proc. Intern. Clay Conf., pp. 137–145. Stockholm (1963).
EGGER, K.: Zur Oxydation natürlicher Magnetite. Schweiz. Mineral. Petrogr. Mitt. **43**, 493–497 (1963).
ENGELHARDT, W. VON, GOLDSCHMIDT, H.: Ein Tonmineral der Kaolinit-Halloysitgruppe von Provins (Frankreich). Heidelb. Beitr. Mineral. Petrogr. **4**, 319–324 (1954).
ERIKSSON, E.: Problems of Heat Flow in d.t.a. Kungl. Lantbrukshögskolans Ann. **19**, 128–143 (1952) and **20**, 117–123 (1953).
FAUST, G. T.: Thermal analysis of quartz and its use in calibration in thermal analysis studies. Am. Mineralogist **33**, 337–345 (1948).
FAUST, G. T.: Thermal analysis studies on carbonates: Aragonite and Calcite. Am. Mineralogist **35**, 207–225 (1950).
FAUST, G. T.: Thermal analysis and X-ray studies of sauconite and of some Zn-minerals of the same paragenetic association. Am. Mineralogist **36**, 795–822 (1951).
FAUST, G. T.: Huntite, a new mineral. Am. Mineralogist **38**, 4–24 (1953).
FENNER, C. N.: The stability relations of the silica minerals. Am. J. Sci. 4th Ser. **36**, 331–384 (1913).
FISCHER, W.: Zum Problem der Achatgenese. Neues Jahrb. Mineral. Abh. **86**, 367–392 (1954).
FLÖRKE, O. W.: Strukturanomalien bei Tridymit und Cristobalit. Ber. Deut. Keram. Ges. **32**, 369–381 (1955).
FLÖRKE, O. W.: Untersuchungen an feinkristallinem Quarz. Schweiz. Mineral. Petrogr. Mitt. **41**, 311–324 (1961).
FLÖRKE, O. W.: Untersuchungen an amorphem und mikrokristallinem SiO_2. Chemie der Erde **22**, 91–110 (1962).

FÖLDVARI-VOGL, M.: The Role of DTA in Mineralogy and Geological Prospecting. Acta Geologica Budapest **5**, 3–102 (1958).
FOMINYKH, V. G., GLUKHIKH, I. I.: Magnetic properties of Titanomagnetite from Ural Titanomagnetite deposits (russ., engl. abstr.). Dokl. Akad. Nauk. SSSR **180**, 710–712 (1958).
FRANZ, E. D.: Stöchiometrischer Kupferkies, stabilisiert durch Substitution von Schwefel durch Selen. Neues Jahrb. Mineral. Monatsh. **1971**, 11–18 (Jan. 1971).
FREDERICKSON, A. J.: D.t.a. curves of siderite. Am. Mineralogist **33**, 372–375 (1948).
FREEMAN, A. G.: The dehydroxylation behaviour of amphiboles. Mineral. Mag. **35**, 953–957 (1966).
FRONDEL, C., BAUER, L. H.: Kutnahorite, a Mn-dolomite. Am. Mineralogist **40**, 748–760 (1955).
GARN, P. D.: Thermoanalytical methods of investigation. New York. Academic Press 1955.
GILLERY, F. H.: Some studies of the low and high forms of SiO_2 and $AlPO_4$ Dff. Nav. Res. Rep. No. **58** (1954).
GRAF, D. L.: Preliminary report on the variations in d.t.a. curves of low-iron dolomites. Am. Mineralogist **37**, 1–27 (1951).
GRANGE, M. H.: Critère pour l'identification d'eau zeolitique dans les hydrates en analyse thermique différentielle et en thermogravimetrie. Compt. Rend. **259**, 3277–3280 (1964).
GRIM, R. E.: D.t.a. curves of prepared mixtures of clay minerals. Am. Mineralogist **32**, 493–501 (1947).
GRIM, R. E., BRADLEY, W. F.: Rehydration and dehydration of the clay minerals. Am. Mineralogist **33**, 50–59 (1948).
GRIM, R. E., JOHNS, W. D., JR.: Reactions accompanying the firing of brick. J. Am. Ceram. Soc. **34**, 71–76 (1951).
GRIM, R. E., ROWLAND, R. A.: D.T.A. of clay minerals and other hydrous minerals. P. I. Am. Mineralogist **27**, 746–761, P. II, 801–818 (1942).
GRIM, R. E., ROWLAND, R. A.: D.t.a. of clays and shales, a control and prospecting method. J. Am. Ceram. Soc. **27**, 65–76 (1944).
GROSS, S.: Determination of spurrite, associated with calcite, by means of DTA. Israel J. Chem. **9**, 601–605 (1971).
GRUBB, P. L. C., HANNAFORD, P.: Magnetism in cassiterite. Mineral. Deps. **2**, 148–171 (1966).
GUTT, W., SMITH, M. A.: An unknown α-phase of $CaSO_4$. Trans. Brit. Ceram. Soc. **66**, 337–345 (1967).
HANSEN, J. W.: Zur Geologie, Petrographie und Geochemie der Bündnerschiefer-Serien zwischen Nufenenpass (Schweiz) und Cascata Toce (Italia), Schweiz. Mineral. Petrogr. Mitt. **52**, 109–153 (1972).
HARADA, K., TOMITA, K.: A Sodian Stilbite from Inogajo, Japan. Am. Mineralogist **52**, 1438–1450 (1967).
HAUL, R. A. W., HEYSTEK, H.: D.t.a. of dolomite decomposition. Am. Mineralogist **37**, 166–179 (1952).
HAUSEN, D. M.: Schoderite, a new phosphovanadate mineral from Nevada. Am. Mineralogist **47**, 637–648 (1962).
HAYASHI, H., KORSHI, K., SAKABE, H.: Structural changes of fibrous minerals-asbestos, sepiolite and palygorskite- on heat treatment and their effect on toxicity to the cells. Proc. Int. Clay Conf. Tokyo, vol. 1, 903–913 (1969).
HEFLIK, W., ZABINSKI, W.: A chromian hydrogrossular from Jordanow, Lower Silesia, Poland. Mineral. Mag. **37**, 241–243 (1969).
HEIDE, K.: Thermische Untersuchungen an Salzmineralen. I. Untersuchungen am Ascharit. Chemie der Erde **22**, 180–221 (1962).

HEIDE, K.: Die Phasenanalyse bei dem thermischen Abbau der Kristallhydrate. Naturwissenschaften **52**, 183–184 (1965).
HEIDE, K.: Thermische Untersuchungen an Salzmineralen II. Chemie der Erde **24**, 94–111 (1965a).
HEIDE, K.: Thermische Untersuchungen an Salzmineralen III. Chemie der Erde **24**, 279–302 (1965b).
HEIDE, K.: Thermische Untersuchungen an Salzmineralen IV. Chemie der Erde **25**, 237–252 (1966).
HEIDE, K.: Die Verwitterung des Kieserits. Chemie der Erde **26**, 133–139 (1967).
HEIDE, K.: Zur Publikation von thermoanalytischen Daten. Chemie der Erde **27**, 190–192 (1968).
HEIDE, K.: Zum Mechanismus der Umbildungsvorgänge in Salzgesteinen. Chemie der Erde **27**, 353–368 (1968).
HEIDE, K.: Thermochemische und kinetische Untersuchungen der endothermen Umbildungsreaktionen des Epsomits ($MgSO_4 \cdot 7H_2O$). J. Thermal Analysis **1**, 183–194 (1969).
HEIDE, K.: Temperaturstandards für die dynamische thermische Analyse. Silikattechnik **3**, 97–98 (1973).
HEIDE, K., BRÜCKNER, U.: Grundlagen zur Phasenanalyse von Salzgesteinen. 1. Die binären Teilsysteme des anhydritischen Hartsalzes KCl-NaCl und KCl-CaSO$_4$. Chemie der Erde **26**, 235–255 (1967).
HEIMANN, K.: Polymorphe Kieselsäurephasen im System $(NaPO_3)_x$-Na_2SiO_3. Glastechn. Ber. **43**, 183–190 (1970).
HENDRICKS, S. B., ALEXANDER, L. T.: Minerals present in soil colloids: I Description and methods for identification. Soil Sci. **48**, 257–271 (1939).
HILLER, J. E., PROBSTHAIN, K.: Eine Apparatur für die DTA von Sulfiden. Erzmetall **8**, 257–267 (1955).
HILLER, J. E., PROBSTHAIN, K.: Thermische und röntgenographische Untersuchungen am Kupferkies. Z. Kristallogr. **108**, 108–129 (1956).
HOCHSTRASSER, K., FEITKNECHT, W.: Thermoanalytische Untersuchungen an feinstdispersen Varietäten der Birnessit- und Kryptomelan-Gruppe. Thermal Anal. vol. **2**: Proc. III. ICTA, Davos 1971, 79–89 (1972).
HODENBERG, R. VON, KÜHN, R., ROSSKOPF, F.: Die chemische Zusammensetzung des Löweit. Kali und Steinsalz **5**, 178–189 (1969).
HOFMANN, F., PETERS, TJ.: Kaolinitische Mergel unter der Molassebasis im Rheinfallgebiet. Schweiz. Mineral. Petrogr. Mitt. **42**, 349–358 (1962).
HOFMANN, U., ROTHE, A.: Über die Veränderung des Quarzes beim trockenen Mahlen. Z. Anorg. Allg. Chemie **357**, 196–201 (1968).
HOSS, H.: Untersuchungen über die Petrographie kulmischer Kieselschiefer. Heidelb. Beitr. Mineral. Petr. **6**, 59–88 (1957).
HUNSINGER, W.: Thermoelektrische Temperaturmeßeinrichtung für hohe Genauigkeitsanforderungen, insbesondere für thermische Analysen. Metallkunde **44**, 261–264 (1953).
HUNZIKER, J. C.: Zur Geologie und Geochemie des Gebietes zwischen Valle Antigorio und Valle di Campo (Tessin). Schweiz. Mineral. Petrogr. Mitt. **46**, 473–552 (1966).
HURLBUT, C. S., JR., ARISTARAIN, L. F.: Rivadavite, a new borate from Argentina. Am. Mineralogist **52**, 326–335 (1967a).
HURLBUT, C. S., JR., ARISTARAIN, L. F.: Ezcurrite, a restudy. Am. Mineralogist **52**, 1048–1059 (1967b).
HURLBUT, C. S., JR., ARISTARAIN, L. F.: Bermanite, and its occurrence in Cordoba, Argentina. Am. Mineralogist **53**, 416–431 (1968a).

HURLBUT, C. S., JR., ARISTARAIN, L. F.: Beusite, a new mineral from Argentina, and the graftonite-beusite series. Am. Mineralogist **53**, 1799–1815 (1968 b).
IJIMA, A., HARADA, K.: Authigenic zeolites in zeolitic palagonite tuffs on Oahu, Hawai. Am. Mineralogist **54**, 182–197 (1969).
IMAI, N., OTSUKA, R., KASHIDE, H., HAYASHI, H.: Dehydration of palygorskite and sepiolite from the Kizum district, Central Japan. Proc. Int. Clay Conf. Tokyo, vol. 1, 99–108 (1969).
JÄGER, E., SCHILLING, S.: Vorläufiger Bericht über DTA-Untersuchungen an Wölsendorfer Fluorit. Schweiz. Mineral. Petrogr. Mitt. **36**, 599–603 (1956).
JASMUND, K.: Die silicatischen Tonminerale. 2. Aufl. Weinheim: Verlag Chemie 1955.
JENNI, H. P.: Über das Pickeringit-Vorkommen von Jutschi (Kt. Uri). Schweiz. Mineral. Petrogr. Mitt. **50**, 276–290 (1970).
KASHKAI, M.-A., BABAEV, I. A.: Thermal investigations on alunite and its mixtures with quartz and dickite. Mineral. Mag. **37**, 128–134 (1969).
KAUTZ, K.: Sedimentpetrographische Untersuchungen zur Diagenese in Sandsteinen der marinen Unterkreide Norddeutschlands. Beitr. Mineral. Petrogr. **9**, 423–461 (1964).
KEATTCH, C.: An introduction to thermogravimetry. London: Heyden & Sons 1969.
KEITH, M. L., TUTTLE, O. F.: Significance of variation in the high-low inversion of quartz. Bowen Vol., Amer. J. Sci. **1952**, 208–280.
KELLER, P.: Eigenschaften von $(Cl,F,OH)_{<2}Fe_8(O,OH)_{16}$ und Akaganeit. Neues Jahrb. Mineral. Abh. **113**, 29–49 (1970).
KELLY, W. C.: Application of d.t.a. to identification of the natural hydrous ferric oxides. Am. Mineralogist **41**, 353–355 (1956).
KERR, P. F., KULP, J. L.: Multiple Differential Thermal Analysis. Am. Mineralogist **33**, 387–419 (1948).
KEUSEN, H. R., BÜRKI, H.: Thomsonit und andere Faserzeolithe als Kluftmineral in Begleitgesteinen der Ultrabasite vom Geisspfadpass im Binntal. Schweiz. Mineral. Petrogr. Mitt. **49**, 577–584 (1969).
KIRSCH, H.: Zur Kristallchemie der Magnetitschutzschichten in den Stahlrohren von Hochdruckdampf-Kraftwerken. Arch. f. Eisenhüttenwesen **36**, 603–608 (1965).
KISSINGER, H. E.: Reaction kinetics in d.t.a. Analyt. Chemistry **29**, 1702–1706 (1957).
KNELLER, W. A., KRIEGE, H. F., SAXER, E. L., WILBRAND, J. T., ROHRBACHER, T. J.: The properties and recognition of deleterious cherts which occur in aggregates used by Ohio concrete products. Res. Found. Univ. Toledo/Ohio, Aggr. Res. Group Geol. Dept. (1968).
KNOKE, R.: Zur Frage der Entstehung der Kieselgallenschiefer. Contrib. Mineral. Petrol. **23**, 236–243 (1969).
KODAMA, H.: The nature of the component layers of rectorite. Am. Mineralogist **51**, 1035–1055 (1966).
KÖHLER, A., WIEDEN, P.: Vorläufige Ergebnisse in der Feldspatgruppe mittels der DTA. Neues Jahrb. Mineral. Monatsh. **1954**, 249–252.
KOHLS, D. W., RODDA, J. L.: Iowaite, a new hydrous Mg-hydroxide-ferric oxychloride from the Precambrian of Iowa. Am. Mineralogist **52**, 1261–1271 (1967).
KOHN, B. P.: Identification of New Zealand tephra layers by emission spectrographic analysis of their titanomagnetites. Lithos **3**, 361–368 (1970).
KOIZUMI, M.: The d.t.a.-curves and the dehydration curves of zeolites. Mineral. J. (Japan) **1**, 36–47 (1953).

KOIZUMI, M., ROY, R.: Zeolite studies, I. Synthesis and stability of the Ca-zeolites. J. Geol. **68**, 41–53 (1960).
KOPP, O. C., KERR, P. F.: DTA of Sulfides and Arsenides. Am. Mineralogist **42**, 445–454 (1957).
KOPP, O. C., KERR, P. F.: DTA of Sphalerite. Am. Mineralogist **43**, 732–748 (1958a).
KOPP, O. C., KERR, P. F.: DTA of Pyrite and Marcasite. Am. Mineralogist **43**, 1079–1097 (1958b).
KORITNIG, S.: Einschlüsse in Suttroper Quarzen. Beitr. Mineral. Petrogr. **8**, 21–27 (1961).
KRACEK, F. C.: Phase relations in the system sulfur-silver and the transitions in silver sulfide. Am. Geophys. Union Trans. **27**, 364–373 (1946).
KRESTEN, P.: Die Genese der Migmatite von Lammholmen, Västervik, SE-Schweden. Stockh. Contrib. Geol. **23**, 91–125 (1971a).
KRESTEN, P.: Die Hoch-Tiefquarz-Inversion als geologisches Thermometer/Barometer. Linseis J. **1**, 6–8 (1971b).
KÜHNEL, R. A., VAN HILTEN, D., ROORDA, H. J.: The crystallinity of minerals in alteration profiles: an example on goethite in laterite profiles. Delft Progr. Rep., Series E.: Geosciences **1**, 1, 1–32 (1974).
KULLERUD, G.: Differential Thermal Analysis. Carnegie Inst. Wash. Yearbook **58**, 161–163 (1959).
KULP, J. L., KENT, P., KERR, P. F.: Thermal study of the Ca-Mg-Fe carbonate minerals. Am. Mineralogist **36**, 643–670 (1951).
KULP, J. L., KERR, P. F.: Multiple d.t.a. Am. Mineralogist **33**, 387–420 (1948).
KULP, J. L., KERR, P. F.: Improved d.t.a. apparatus. Am. Mineralogist **34**, 839–845 (1949).
KULP, L., PERFETTI, J. N.: Thermal study of some Mn-oxide minerals. Mineral. Mag. **29**, 239–251 (1950).
KULP, J. L., TRITES, A. F.: D.t.a. of the natural hydrous ferric oxides. Am. Mineralogist **36**, 23–44 (1951).
KURZWEIL, H.: Sedimentpetrologische Untersuchungen an den jungtertiären Tonmergelserien der Molassezone Oberösterreichs. Tschermaks Mineral. Petrogr. Mitt. **20**, 196–215 (1973).
LAMEYRE, J., LÉVY, C., MERGOIL, J.: Etude par analyse thermique différentielle de quartz de leucogranites et de schistes cristallins du Massif Central français. Bull. Soc. Franc. Minéral. Cryst. **91**, 172–181 (1968).
LANGER, A. M., KERR, P. F.: Evaluation of kaolinite and quartz d.t.a. curves with a new high temperature cell. Am. Mineralogist **52**, 508–523 (1967).
LANGER, H. G., GOHLKE, R. S.: Mass Spectrometric Thermal Analysis (MTA). Analyt. Chem., **35**, 1301–1302 (1963).
LANGER, H. G., GOHLKE, R. S., SMITH, D. H.: Mass Spectrometric Differential Thermal Analysis. Analyt. Chem. **37**, 433–434 (1965).
LAPHAM, D. L.: Structural and chemical variation in chromium chlorite. Am. Mineralogist **43**, 921–956 (1958).
LE CHATELIER, H.: De l'action de la chaleur sur les argiles. Bull. Soc. Franç. Minéral. Cryst. **10**, 204–211 (1887).
LEHMANN, H.: Neuere Entwicklungen auf dem Gebiete der DTA. Deut. Keram. Ges. Ber. **32**, 172–175 (1955).
LEHMANN, H., DAS, S. S., PAETSCH, H. H.: Die Differentialthermoanalyse. Tonind.-Ztg. Keram. Rundsch. 1. Beiheft (1953).
LEHMANN, H., HOLLAND, H.: Der Prozeß der Gips-Transformation und seine Entwässerungsprodukte durch Erhitzen. Tonind.-Ztg. **90**, 2–21 (1966).
LÉVY, C.: Analyse thermique différentielle des minéraux sulfurés. Bull. Soc. Franç. Minéral. Cryst. **81**, 29–34 (1958).

Lewis, J. F.: Chemical composition and physical properties of magnetite from the ejected plutonic blocks of the Soufrière volcano, St. Vincent, West Indies. Am. Mineralogist **55**, 793–807 (1970).
Linseis, M.: Eine verbesserte DTA-Apparatur und deren Anwendung. Sprechsaal **85**, 423–427 (1952).
Lippmann, F.: Über einen Keuperton von Zaisersweiher b. Maulbronn. Heidelb. Beitr. Mineral. Petrogr. **4**, 130–134 (1954).
Lippmann, F.: Über eine Apparatur zur DTA. Keram.-Z. (Lübeck) **11**, 475, 524, 570 (1959).
Ljunggren, P. H.: Determination of mineralogical transformations of gypsum by DTA. J. Am. Ceram. Soc. **43**, 227 (1960).
Lugscheider, W.: Die DTA zur Bestimmung der Schmelzenthalpien von Metallen und Legierungen bei Temperaturen bis 1200° C. Ber. Bunsen-Ges. **71**, 228–235 (1967).
Mackenzie, R. C.: Investigations on cold-precipitated ferric oxide and its origin in clays. Problems of Clay and Laterite Genesis, p.65–75. New York: Am. Inst. Mining. Ing. 1952.
Mackenzie, R. C. (Ed.): The Differential Thermal Investigation of Clays. Mineral. Soc. (Clay Mineral. Group), London (1957).
Mackenzie, R. C.: Simple phyllosilicates based on gibbsite- and brucite-like sheets. Chapter 18. In: Mackenzie, R. C.: Differential Thermal Anal., Vol. 1, pp. 497–537 (1970).
Mackenzie, R. C. (Ed.): Differential Thermal Analysis, Vol. 1. London: Academic Press 1970.
Mackenzie, R. C. (Ed.): Differential Thermal Analysis, Vol. 2. London: Academic Press 1972.
Mackenzie, R. C., Milne, A.: The effect of grinding on micas. I. Muscovite. Mineral. Mag. **30**, 178–185 (1953).
Marel, H. W., van der: Quantitative d.t.a. of clay and other minerals. Am. Mineralogist **41**, 222–244 (1956).
Mason, B., Sand, L. B.: Clinoptilolite from Patagonia. Am. Mineralogist **45**, 341–350 (1960).
Maurel, C.: Types de réactions d'oxydation observés au cours de l'analyse thermique différentielle, dans l'air, de minéraux sulfurés et arséniés de Fe, Co, Ni, Cu, Zn, Ag et Pb. Bull. Soc. Franç. Minéral. Cryst. **87**, 377–385 (1964).
McAdie, H. G.: Recommendations for Reporting Thermal Analysis Data. Analyt. Chem. **39**, No. 4, 543 (1967).
McAdie, H. G., Wiedemann, H. G.: Empfehlungen für die Mitteilungen von Daten thermoanalytischer Untersuchungen. Analyt. Chem. **231**, 36–38 (1967).
McDougall, D. J.: A "lattice defect-free energy" approach to replacement processes in ore deposition. Econ. Geol. **63**, 671–681 (1968).
McLaughlin, R. J. W.: Thermal Techniques. In: Zussman, J. (Ed.): Physical Methods in Determinative Mineralogy. London: Academic Press, Inc. 1967.
Merkle, A. B., Slaughter, M.: Determination and refinement of the structure of heulandite. Am. Mineralogist **53**, 1120–1138 (1968).
Midgley, H. G.: A serpentine mineral from Kennack Cove, Lizard, Cornwall. Mineral. Mag. **29**, 526–530 (1951).
Moore, G. S. M., Rose, H. E.: The Structure of Powdered Quartz. Nature **242**, No. 5394, 187–190 (1973).
Müller, Germ.: Sediment-Petrologie, Teil I: Methoden der Sediment-Untersuchung. Stuttgart: E. Schweizerbart'sche Verlagsbuchhandlung 1964.
Müller-Vonmoos, M., Schindler, C.: Palygorskit im helvetischen Kieselkalk des Bürgenstocks. Schweiz. Mineral. Petrogr. Mitt. **53**, 3, 395–403 (1973).

MURPHY, C. B., HILL, J. A., SCHACHER, G. P.: Differential thermal analysis and simultaneous gas analysis. Analyt. Chem. **32**, 1374 (1960).
NAUMANN, A. W., DRESHER, W. H.: The influence of sample texture on chrysotile dehydroxylation. Am. Mineralogist **51**, 1200–1211 (1966).
NORTON, F. H.: Critical study of d.t.a. for identification of the clay minerals. J. Am. Ceram. Soc. **22**, 54–63 (1939a).
NORTON, F. H.: Hydrothermal formation of clay minerals in the laboratory. Am. Mineralogist **24**, 1–17 (1939b).
ORCEL, J.: D.t.a. in the determination of the consistents of clays, laterites, and bauxites. Congr. Intern. d. Mines, Met. Geol. Applic., VII, Session. pp. 359–373, Paris (1935).
ORCEL, J., CAILLÈRE, S.: L'analyse thermique différentielle des argiles à montmorillonite (bentonite). Compt. Rnd. **1933**, 774–777.
ORCEL, J., CAILLÈRE, S., HÈNIN, S.: Nouvel essai de classification des chlorites. Mineral. Mag. **29**, 329–340 (1951).
PAULIK, F., GAL, S., ERDEY, L.: Pyrit-Gehalte in Bauxiten. Anal. Chim. Acta **29**, 381–394 (1963).
PAULIK, F., PAULIK, J.: Investigations under quasi-isothermal and quasi-isobaric conditions by means of the derivatograph. Thermal Anal. **5**, 253–270 (1973).
PAULIK, F., PAULIK, J., ERDEY, L.: Derivatography, a complex Method in Thermal Analysis. Talanta (London) **13**, 1405–1430 (1966a).
PAULIK, F., PAULIK, J., ERDEY, L.: Kombinierte derivatographische und thermogasanalytische Untersuchungen. Mikrochim. Acta (Wien) **1966b**, 886–893.
PAULIK, F., PAULIK, J., ERDEY, L.: Derivative Dilatometrie. Mikrochim. Acta (Wien) **1966c**, 893–902.
PÉCSINÉ-DONATH, E.: DTA-Untersuchung über die thermische Zersetzung der Zeolithe. Földt. Közl., Tonmin.-Bd. **93**, 32–39 (1963).
PENG, C. J.: Thermal study of the natrolite group. Am. Mineralogist **40**, 834–856 (1955).
PETERS, TJ.: Differentialthermoanalyse von Vesuvian. Schweiz. Mineral. Petrogr. Mitt. **41**, 325–334 (1961).
PETERS, TJ.: Tonmineralogische Untersuchungen an Opalinustonen und einem Oxfordienprofil im Schweizer Jura. Schweiz. Mineral. Petrogr. Mitt. **42**, 359–380 (1962).
PETERS, TJ.: Mineralogie und Petrographie des Totalpserpentins bei Davos. Schweiz. Mineral. Petrogr. Mitt. **43**, 529–685 (1963).
PHILLIPS, W. R.: A differential thermal study of the chlorites. Mineral. Mag. **33**, 404–414 (1954).
PIÈCE, R.: Analyse thermique différentielle et thermogravimétrie simultanées du gypse et de ses produits de déshydratation. Schweiz. Mineral. Petrogr. Mitt. **41**, 303–310 (1961).
PLAS, L. VAN DER: Personal communication.
PLAS, L. VAN DER, HÜGI, TH.: A Ferrian Sodium-Amphibole from Vals, Switzerland. Schweiz. Mineral. Petrogr. Mitt. **41**, 371–393 (1961).
POSNJAK, E., ALLEN, E. T., MERWIN, H. E.: The sulphides of copper. Econ. Geol. **10**, 491–535 (1915).
POWELL, D. A.: Setting of gypsum plaster. Nature **178**, 428–429 (1956).
PUSZTASZERI, L.: Étude petrographique du massif du Chenaillet (Hautes Alpes, France). Schweiz. Mineral. Petrogr. Mitt. **49**, 425–466 (1969).
RAMDOHR, P., STRUNZ, H.: Klockmanns Lehrbuch der Mineralogie. 15. Aufl. Stuttgart: F. Enke-Verlag 1967.
REDDICK, K. L.: Personal communication.

ROKOSZ, A., PAULIK, J., PAULIK, F., ERDEY, L.: Beiträge zur quantitativen Differentialthermoanalyse. Acta Chim. Acad. Sci. Hung. Tom. **56**, 3, 221–227 (1968).

ROSEBOOM, E., JR.: An investigation of the system Cu-S and some natural copper sulfides between 25° and 700° C. Econ Geol. **61**, 641–672 (1966).

ROSENHAUER, B., SCHWAB, R. G., JABLONSKI, K. H.: High pressure DTA of pyroxenes up to 1400° C (1974, in press).

ROSS, G. J.: Structural decomposition of an orthochlorite during its acid dissolution. Can. Mineralogist **9**, 522–530 (1968).

ROSS, G. J., KODAMA, H.: Properties of a synthetic Mg-Al-Carbonate-Hydroxide and its relationship to Mg-Al double hydroxide, manasseite and hydrotalcite. Am. Mineralogist **52**, 1036–1047 (1967).

ROWLAND, R. A., BECK, C. W.: Determination of small quantities of dolomites by d.t.a. Am. Mineralogist **37**, 76–82 (1952).

ROWLAND, R. A., JONAS, E. C.: Variations in d.t.a. curves of siderites. Am. Mineralogist **34**, 550–558 (1949).

SABATIER, G.: Analyse thermique différentielle de quelques sulfures. Bull. Soc. Franç. Minéral. Cryst. **79**, 172–174 (1956).

SABATIER, G.: Chaleurs de transitions des formes de basse température du quartz, de la tridymite et de la cristobalite. Bull. Soc. Franc. Minéral. Cryst. **80**, 444–449 (1957).

SADANAGA, R., SUENO, S.: X-ray study of the α-β transition of Ag_2S. Mineral. J. (Japan) **5**, 124–148 (1967).

SAITO, M., KAKITANI, S., UMEGAKI, Y.: Thermal studies of some minerals, No. 3: On the transformation of antigorite in air. J. Sci. Hiroshima Univ., Series C (Geology and Mineralogy), **6**, No. 4, 331–342 (1972).

SALGER, M.: Verwitterung und Bodenbildung auf diluvialen Schotterterrassen. Heidelb. Beitr. Mineral. Petrogr. **4**, 288–318 (1954).

SAND, L. B., BATES, T. B.: Quantitative analysis of endellite, halloysite and kaolinite by d.t.a. Am. Mineralogist **38**, 271–278 (1953).

SCHMIDT, E. R., VERMAAS, F. H. S.: DTA and cell dimensions of some natural magnetites. Am. Mineralogist **40**, 422–440 (1955).

SCHMITZ, H., GÄRTNER, H. R. VON: Die organische Substanz des Posidonienschiefers. Erdöl und Kohle **16** (1963).

SCHÜLLER, K. H.: Mineralogische und chemische Untersuchungen am Göpfersgrüner Speckstein. Neues Jahrb. Mineral. Monatsh. 1968, 363–376.

SCHULTZE, D.: Differentialthermoanalyse. Weinheim: Verlag Chemie 1969.

SCHWAB, R. G.: Die Phasenbeziehungen der Pyroxene im System $CaMgSi_2O_6$-$MgSiO_3$-$FeSiO_3$. Fortschr. Mineralogie **46**, 188–273 (1969).

SCHWAB, R. G., JABLONSKI, K. H.: Der Polymorphismus der Pigeonite. Fortschr. Mineralogie **50**, 223–263 (1973).

SCHWANDER, H., HUNZIKER, J., STERN, W.: Zur Mineralchemie von Hellglimmern in den Tessiner Alpen. Schweiz. Mineral. Petrogr. Mitt. **48**, 357–390 (1968).

SCIFAC, D.T.A. Data Index, compiled by R. C. MACKENZIE. London: Cheaver-Hume Press 1962.

SIDDIQUI, M. K. H.: Palygorskite clays from Andhra Pradesh, India. Clay Mineral. **7**, 120–123 (1967).

SKINNER, B. J.: The system Ag-Cu-S. Econ. Geol. **61**, 1–26 (1966).

SMITH, J. W., JOHNSON, D. R., MÜLLER-VONMOOS, M.: Dolomite for determining atmosphere control in thermal analysis. Thermochim. Acta **8**, 45–56 (1974).

SMOTHERS, W. J., CHIANG, Y.: Handbook of Differential Thermal Analysis. New York: Chem. Publ. Co. Inc. 1966.

SMYKATZ-KLOSS, W.: Differential-Thermo-Analyse von einigen Karbonat-Mineralen. Beitr. Mineral. Petrogr. **9**, 481–502 (1964).

Smykatz-Kloss, W.: Sedimentpetrographische und geochemische Untersuchungen an Karbonatgesteinen des Zechsteins. Teil I: Methodischer Teil. Contrib. Mineral. Petrol. **13**, 207–231 (1966).

Smykatz-Kloss, W.: Über die Möglichkeit der halbquantitativen Mineralbestimmung mit der DTA ohne Flächenintegration. Contrib. Mineral. Petrogr. **16**, 274–278 (1967a).

Smykatz-Kloss, W.: DTA einiger bei sehr hohen Temperaturen zerfallender Karbonatminerale. Contrib. Mineral. Petrol. **16**, 279–283 (1967b).

Smykatz-Kloss, W.: Über die Genese der Quarze von Dietlingen in Baden und von Suttrop in Westfalen. Neues Jahrb. Mineral. Monatsh. **1969**, 563–567.

Smykatz-Kloss, W.: Die Hoch-Tiefquarz-Inversion als petrologisches Hilfsmittel. Contrib. Mineral. Petrol. **26**, 20–41 (1970).

Smykatz-Kloss, W.: Petrologische Anwendung der Inversionstemperatur-Bestimmung von Quarzen. Thermal Anal., Vol. 3; Proc. III. ICTA, Davos 1971, 637–648.

Smykatz-Kloss, W.: Das Hoch-Tief-Umwandlungsverhalten mikrokristalliner Quarze. Contrib. Mineral. Petrol. **36**, 1–18 (1972a).

Smykatz-Kloss, W.: Einkieselungen und Verkieselungen in einigen Gesteinen des Schwarzwaldes. Oberrhein. Geol. Abh. **21**, 75–85 (1972b).

Smykatz-Kloss, W.: The determination of the degree of (dis-)order of kaolinites by means of DTA. Chemie der Erde (1974, in press).

Smykatz-Kloss, W., Schultz, R.: Zusammenhang zwischen Fehlordnungsgrad und Bildungstemperatur bei synthetischen Cristobaliten. Contrib. Mineral. Petrol. **45**, 15–25 (1974).

Sosman, R. B.: The Phases of Silica. N. Brunswick, N.J.: Rutgers Univ. Press 1965.

Speil, S., Berkelhamer, L. H., Pask, J. A., Davis, B.: D.t.a. Its application to clays and other aluminous materials. U.S. Dept. Inst., Bur. Mines. Techn. Paper 664, 81 pp. (1945).

Spotts, J. H.: X-ray studies and d.t.a. of some coastal limestones and associated carbonates of W-Australia. Thesis, Univ. of W-Australia, Perth. 23 pp. 1952.

Stärk, H., Jun.: Meßgenaue und meßempfindliche Registriereinrichtungen. Hartmann und Braun-Druckschrift 3033.

Stalder, H. A.: Petrographische und mineralogische Untersuchungen im Grimselgebiet. Schweiz. Mineral. Petrogr. Mitt. **44**, 187–399 (1964).

Strunz, H.: Mineralogische Tabellen. 5. Aufl. Leipzig: Akad. Verl.-Ges. 1970.

Sudo, T., Shimoda, S.: Interstratified Phyllosilicates, Chapter 19. In: Mackenzie, R. C. (Ed.): Differential Thermal Anal., Vol. 1, pp. 539–551 (1970).

Sudo, T., Shimoda, S., Nishigaki, S., Aoki, M.: Energy changes in dehydration processes of clay minerals. Clay Mineral. **7**, 33–42 (1966).

Suhr, N.: Concerning the Ag_2S-Cu_2S system. Econ. Geol. **50**, 347–350 (1955).

Sveshnikovr, V. N., Kuznetsov, V. G.: Structural relationships between zeolites and natural kaolin and their transformations on heating (Russian). Izv. Akad. Nauk USSR **1946**, 25–36.

Takeuchi, T., Takahashi, I., Abe, H.: Wall rock alteration and genesis of sulphur and iron sulphide deposits in N-Japan. The Sci. Rep. Tohuku-Univ., Sendai, Japan, Vol. IX, pp. 371–484 (1966).

Tets, A. van, Wiedemann, H.-G.: Differenztemperatur-Kurven bei Schmelz-, Unterkühlungs- und Erstarrungs-Prozessen. Naturwissenschaften **54**, H. 24 (1967).

Thiel, R.: Zum System α-FeOOH-α-AlOOH. Z. Anorg. Allg. Chem. **326**, 70–78 (1963).

Tobschall, H.-J.: Eine Subfaziesfolge der Grünschieferfazies in den Mittleren Cévennen (Dép. Ardèche) mit Pyrophyllit aufweisenden Paragenesen. Contrib. Mineral. Petrol. **24**, 76–91 (1969).

TOBSCHALL, H.-J.: Zur Genese der Migmatite des Beaume-Tales (Mittl. Cévennen, Dép. Ardèche). Contrib. Mineral. Petrol. **32**, 93–111 (1971).

TRDLIČKA, Z.: Beitrag zur DTA der Gang-Mangancalcite und -Rhodochrosite der CSSR (tchech., German summary) Acta Univ. Carolina (Prag), Geologica, **1964**, 159–167.

TRDLIČKA, Z.: DT- und chemische Analysen der Dolomite und Ankerite aus der CSSR. Acta Univ. Carolina (Prag), Geologica **1966**, 129–136.

TROCHIM, H. D.: Chlorit-Minerale, In: TRÖGER, P., BRAITSCH, O.: Optische Bestimmung der gesteinsbildenden Minerale, Teil II: Textband. Stuttgart: Schweizerbart'sche Verlagsbuchhandlung 1967.

TSUESUE, A.: Magnesian Kutnahorite from Rynjima mine, Japan. Am. Mineralogist **52**, 1751–1761 (1967).

TURNER, B.: Differential thermal analysis in a Piston-cylinder apparatus. High Temp.-High Press. **5**, 273–277 (1973).

TUTTLE, O. F.: The variable inversion temperature of quartz as a possible geologic thermometer. Am. Mineralogist **34**, 723–730 (1949).

VALKENBURG, A. VAN, RYNDERS, G. F.: Synthetic cuspidine. Am. Mineralogist **43**, 1197–1202 (1958).

VINCENT, E. A., WRIGHT, J. B., CHEVALLIER, R., MATHIEU, S.: Heating experiments on some natural titaniferous magnetites. Mineral. Mag. **31**, 624–655 (1957).

VIVALDI, J. L. M., HACH-ALI, P. F.: Palygorskites and Sepiolites. Chapter 20. In: MACKENZIE, R. C.: Differential Thermal Analysis, Vol. 1, pp. 553–573 (1970).

WALENTA, K.: Grimselit, ein neues K-Na-Uranyl-Karbonat aus dem Grimselgebiet (Oberhasli, Kt. Bern, Schweiz). Schweiz. Mineral. Petrogr. Mitt. **52**, 93–108 (1972).

WAMBEKE, L. VAN: The uranium-bearing mineral bolivarite: new data and a second occurence. Mineral. Mag. **38**, 418–423 (1971).

WARNE, S. ST. J.: The detection and identification of the silica minerals by d.t.a. J. Inst. Fuel. (Aust.) **1970**, 240–242.

WARNE, S. ST. J., BAYLISS, P.: DTA of cerussite. Am. Mineralogist **47**, 1011–1023 (1962).

WATERS, B. H. J.: A study of carbonate minerals by d.t.a. Aust. Mineral. Develop. Labs. Bull. No. **3**, 31–36 (1967).

WATSON, E. S., O'NEILL, M. J., JUSTIN, J., BRENNER, N.: A differential scanning calorimeter for quantitative differential thermal analysis. Analyt. Chem. **36**, 1233–1238 (1964).

WEAVER, CH. E.: A lath shaped non-expanded dioctahedral 2:1 clay mineral. Am. Mineralogist **38**, 279–289 (1953).

WEBB, T. L.: Thesis, Univ. of Pretoria, S-Africa (1958).

WEBB, T. L.: Personal communication.

WEBB, T. L., KRÜGER, J. E.: Carbonates, Chapter 10. In: MACKENZIE, R. C. (Ed.): Differential thermal analysis, Vol. 1, pp. 303–341 (1970).

WEFERS, K.: Gleichzeitige Röntgen- und DTA-Untersuchung fester Stoffe. Ber. Deut. Keram. Ges. **42**, 35–38 (1965).

WEISS, A., KOCH, G., HOFMANN, U.: Zur Kenntnis von Saponit. Ber. Deut. Keram. Ges. **32**, 12–17 (1955).

WEISSE, E.: Ein neues Gerät für die thermische Analyse. Aluminium-Archiv **26**, 1–4 (1939).

WEITZEL, H.: Suszeptibilitäten von Mischkristallen $(Mn, Fe)WO_4$, Wolframit. Neues Jahrb. Mineral. Abh. **113**, 13–28 (1970).

WENDLANDT, W. W.: Thermal methods of analysis. New York: Intersci. Publ. Wiley & Sons 1964.

WETZEL, R.: Chemismus und physikalische Parameter einiger Chlorite aus der Grünschieferfazies. Schweiz. Mineral. Petrogr. Mitt. **53**, 273–298 (1973).

WHITE, W. A.: Allophanes from Lawrence Co., Indiana. Am. Mineralogist **38**, 634–642 (1953).

WIEDEMANN, H. G.: Universelles Meßgerät für gravimetrische Untersuchungen unter veränderlichen Bedingungen. Chemie-Ingenieur-Technik **36**, 1105–1114 (1964).

WIEDEMANN, H. G., TETS, A. VAN: Kurvenverlauf der DTA-Messung bei Schmelz- und Erstarrungsvorgängen. Calorimetrische Eichung der DTA-Apparatur durch Messungen an metallischen Proben. Z. Anal. Chemie **233**, 161–175 (1968).

WILSON, M. J., BAIN, D. C., MITCHELL, W. A.: Saponite from the Dalradian metalimestones of N-E Scotland. Clay Miner. **7**, 343–349 (1968).

WILSON, M. J., BERROW, M. L., MCHARDY, W. J.: Lithiophorite from the Lecht mines, Tomintoul, Banffshire. Mineral. Mag. **37**, 618–623 (1970).

WITTELS, M.: The structural disintegration of some amphiboles. Am. Mineralogist **37**, 28–36 (1952).

WONDRATSCHEK, H.: Über die Vorgänge bei der Entwässerung des Chrysotils. Veröff. Max-Planck-Inst. Silikatforsch. **1957**, 1–5.

WRIGHT, J. B.: The iron-titanium oxides of some Dunedin lavas. Mineral. Mag. **36**, 425–435 (1967).

WYLLIE, P. J., RAYNOR, E. J.: DTA and quenching methods in the system $CaO-CO_2-H_2O$. Am. Mineralogist **50**, 2077–2082 (1965).

WYLLIE, P. J., TUTTLE, O. F.: The quenching technique in non-quenchable systems. A discussion concerning the alleged thermal decomposition of portlandite at high pressures. Am. J. Sci. **261**, 983–988 (1963).

ZEMANN, J.: Kristallchemie, Göschen Bd. 1220/1220a (1966).

Subject Index

a, indirect DTA characteristic for sheet silicates 83f.
acanthite 31
—, DTA curve 27
—, DTA data 29
—, inversion behaviour 116f.
activation energy 21
active reference material 7
agates 134ff., 139, 150ff.
—, channels of diffusion 153
—, DTA data 150
Ag-sulfides, inversion behaviour 116
Al-chlorite 69, 73, 126ff.
—, DTA curve 75
—, DTA data 76
— montmorillonite mixed layer 81
—, PA-curve 77
Al-goethite, DTA data 39
Al-hydroxides 37f., 40
Al-incorporation into goethite 114
allophane 18, 91
—, DTA data 92
alloying, Pt by S 25
alstonite 9
alunite, DTA curve 34
—, DTA data 36
amblygonite 60
—, DTA curve 59
—, DTA data 58
ameghinite 57
amount of sample 8ff., 18, 57
— —, optimal 9
amphiboles 63
amplified initial temperature 4
amplifier 3, 15
anglesite 36
—, DTA curve 34
—, DTA data 36
ankerite 41, 45, 56, 112, 114
—, DTA curve 42
—, DTA data 44

ankerite, PA-curve 46
annabergite, DTA curve 59
—, DTA data 58
—, PA-curve 61
antigorite 78, 127
—, *a*-value 84
—, DTA curve 78
—, DTA data 79
antimonides 25
aphrosiderite 78, 126
—, DTA curve 75
—, DTA data 78
apparatus 3, 10, 15
—, calibration 13
aragonite → calcite transformation 48, 109, 115
—, heat of reaction 20
— , incorporation of Ba and Sr 48, 108ff.
argentite 116
arsenates 58f.
arsenides 25, 27
arsenopyrite, DTA curve 26
—, DTA data 29
artinite, DTA curve and data 54
—, PA-curve 55
ascharite 57
astrakanite, DTA curve 34
—, DTA data 36
attapulgite 79
aurichalcite 114
—, DTA curve and data 49
—, PA-curve 50
authigenic quartz, defects 143, 145f.
— —, inversion temperature 37, 141, 154
— —, —, variations 141
a-values 95
—, sheet silicates 84
azurite, DTA curve 49
—, DTA data 49

Ba-incorporation into aragonite 109
ball jasper, DTA data 151
banded jasper, DTA data 151
basalt jasper, DTA data 151
base line 4, 18
— shift 4, 5
bastnaesite, DTA curve 49
—, DTA data 49
—, PA-curve 50
bauxite 40
—, DTA data 41
bayerite, DTA curve 38
—, DTA data 39
—, PA-curve 40
bean ore 40
benstonite 9
bermanite 62
beustite 62
β-FeOOH 38
biotite 72f.
—, DTA curve and data 74
—, Ti content 73
bitumen 17, 92
blind run 149
bloedite 34
boehmite, disorder, DTA criteria 39
—, DTA curve 38
—, DTA data 39
bolivarite 62
boracite 57
borates 57, 59
borax, DTA curve 59
—, DTA data 58
bornite 28
—, DTA curve 26
—, DTA data 29
bournonite 28
—, DTA curve 26
—, DTA data 29
„brauner Glaskopf" 40
—, DTA data 41
breunnerite 45, 56
—, DTA curve 42
—, DTA data 44
brown ore 40
brucite, DTA curve 38
—, DTA data 39
brugnatellite, DTA curve and data 54
—, PA-curve 55

calcite 9, 12, 54
—, decomposition, heat of reaction 20
—, DTA curve 43

calcite, DTA data 44
—, influence of substitution on decomposition temperature 108f.
—, PA-curve 46
calibration, DTA apparatus and thermocouples 13ff., 36f., 47, 57, 116, 140
—, caloric 20
— curves 13, 17, 132, 149
—, materials 14
caloric calibration 20
— conductivity, combined with DTA 23
calorimetry 23, 36
carbonates 41ff., 94
—, decomposition 9, 12, 17
—, —, heat of reaction 20
—, difficulties in DTA of Fe-Mn $\sim$41f.
—, PA-curves 56
—, structural transformations 47f.
carletonite 41
cassiterites 119
cation exchange capacity 71
ceramic raw materials 17
— sample holder 6, 25, 27
cerussite 41
—, DTA curve 42
—, DTA data 44
—, PA-curve 52
chabasite 88f., 91, 94
—, DTA curve 88
—, DTA data 90
—, PA-curve 89
chain silicates 63
chalcedony 134, 138, 147f.
—, DTA data 150f.
—, secondary in agates 153
chalcocite 28
—, DTA curve 26
—, DTA data 29
—, inversion behaviour 116f.
—, —, dependence on Fe-content 118
chalcogenides, DTA characteristic 25f., 115ff.
—, oxidation behaviour 27
chalcopyrite 28
—, DTA curve 26
—, DTA data 29
chamosite 73, 125
—, DTA curve 75
—, DTA data 76
—, PA-curve 77

chemical composition of minerals, influence on t and ΔT 94, 107ff.
— — —, structural transformations 115ff.
Chile nitre 57
chlorides 31f.
chlorite 67, 69, 73, 78, 84ff.
—, comparison with serpentine data 127
—, decomposition 77
—, dehydration 73f.
—, DTA classification 73, 122ff.
—, —, classes 126
—, — curves 75
—, — data 76
—, interdependence between chemical composition, decomposition temperature and exothermic peak temperature 126
—, — exothermic peak temperature and Fe-content 125f.
— montmorillonite mixed layer 83
—, —, a-value 84
—, —, DTA curve 83
—, preparation 123
—, substitutions, influence on DTA data 75, 77, 123
chromite-magnetite 120
chrysocolla 63
—, DTA curve 63
chrysoprase 147, 152
—, DTA data 150
chrysotile 78, 127
—, a-value 84
—, DTA curve 78
—, DTA data 79
clausthalite 28
clay minerals 40, 64ff., 94, 107, 125, 130, 132
— —, dehydration behaviour 9f., 12, 17f.
— —, —, heat of reaction 20
— —, —, pressure dependence 22
— —, Fe-rich 84
— —, indirect DTA characteristic a 83f.
clays 127
—, "quantitative" DTA 19
claystones 131
clinochlor 78, 126f.
—, DTA curve 75
—, DTA data 76

CO_2, reaction product 9
coal 92
colemanite 57
—, DTA curve 59
—, DTA data 58
combined methods 21ff.
combustion 3, 17, 24
—, organic matter 92
comparison of DTA data 3, 12
contactmetamorphic quartz, t_i-variation 141, 145
cookeite 73, 126
—, DTA curve 75
cooling 137
— rate 11
cooperite 25
corrensite 81f., 84
—, a-value 84
—, DTA curve 83
—, DTA data 82
—, PA-curve 77
covellite 116
Cr-chlorites 75, 77, 123f., 126f.
—, DTA curve 75
Cr-incorporation into chlorite, DTA determination 123
cristobalite, t_i 37, 142, 146, 152
—, low-temperature $\sim$, inversion behaviour 133ff.
—, synthetic, interdependence between t_i and synthesis temperature 135f.
cryolite 17, 31, 140
—, DTA curve and data 32
crystal physical defects, influence on t and ΔT 107
crystallinity, goethites 39f.
crystallization 33
Cu-incorporation into hydrozincite, DTA determination 114
Curie point, Fe-oxides 37
— temperature, magnetites 118ff.
— —, nickel 119
cuspidine 33
Cu-sulfides, distinction by DTA 117
—, DTA curves 26
—, inversion behaviour 28, 116

decomposition 107
—, carbonates 9, 12
—, —, influence of substitutions 108ff.
—, chlorite 72
—, dolomite 111

decomposition, kaolinite 132
— peak, chlorites, influence by substitutions 123
— —, influence by degree of disorder 127
—, serpentine 79
—, sheet silicates 83
—, smectites and vermiculites 71, 129
— temperatures, carbonates 43
— —, dolomite 111 f.
— —, Mg-calcite 108
defect character, crystals 8, 10, 18
— —, influence on decomposition temperatures 107 ff.
degree of crystallization 40
— of disorder 8, 10, 37, 142 f., 145 f., 152
— —, chlorites 127
— —, cristobalites 135 ff.
— —, goethite 38, 39
— —, hydroxides 114 f.
— —, influence on t and ΔT 94
— —, kaolinites 66, 131 f.
— —, serpentines 84
dehydration 2, 6, 9, 12, 17, 33, 94
—, allophanes 91
—, chlorites 73 f.
—, heats of reaction 20
—, hydrated carbonates 50, 52
—, hydroxides 38
—, —, influence by chemical composition 39
—, opal 92
—, partial pressure of H_2O 9
—, pressure dependence 22
—, "quantitative" DTA 18
—, sheet silicates 64
—, silicates 63
—, smectites and vermiculites 71, 129 f.
—, sulfates 33 f.
derivative DTA 23
derivatograph 1, 22 f.
diagenesis 108, 131, 138 ff., 143, 145 f., 153 ff.
diaspore 40
—, DTA curve 38
—, DTA data 39
dickite 66, 131, 133
—, DTA curve 65
—, DTA data 67
differential calorimetry 23
— enthalpy analysis 23

differential thermal analysis see DTA
— thermal gravimetry 23
digenite 116 ff.
dilatometry, combined with DTA 23
dioptase 63
direct DTA characteristics 4, 13
disorder 8, 10, 18, 64, 66, 70, 84, 107, 127
—, boehmite 39
—, cristobalite 134 ff.
—, definition 134
—, DTA criteria 38
—, goethite 38
—, kaolinite 40, 130 ff.
—, origin 143
displacive transformation, cristobalite 133
— —, quartz 140
— —, sulfides 116
dissociation 33
di/tri-octahedral minerals, DTA distinction 128 ff.
djurleite 116 ff.
dolomite 42, 54
—, DTA curve 43, 112
—, DTA data 44
—, Fe-Mn-incorporation, influence on decomposition temperature 111 f.
—, PA-curve 46
drifting of the base line (zero line) 6, 8, 18, 53
—, zeolites 89
ΔT, dependence on chemical composition and degree of disorder 94
—, measurement 93 f.
DTA characteristics 4 f.

earth wax, DTA data 92
Einkieselung 153, 157
elements, DTA characteristics 24 f.
endothermic reactions 2 ff.
enthalpy 21, 23
epidotes 63
epsomite 33, 36
equilibrium temperatures, thermodynamic 4, 7, 13, 19, 21, 33
— —, DTA determination 19
equipment factors 3, 5 ff., 18
erythrite, DTA curve 59
—, DTA data 58
evaporation 3, 65
—, sulfates 33

exactness of measurement 15 f.
— —, improvement by internal standards 16 f.
exothermic effect, disordered structures 38
— peak, criterion of sedimentary chlorite classification 127
— peak temperatures, vermiculites, measure for the di-octahedral part 128
— reactions 3 ff.
— —, influence of pressure variation 22
ezcurrite 57

factors influencing DTA, equipment 3, 5 ff.
— — —, preparative 10
Fe-chlorites 73 f., 124 ff.
—, a-values 84
—, DTA curves 75
—, DTA data 76
—, PA-curve 77
Fe-dolomite 42
—, DTA curves 112
—, DTA determination of Fe-content 111 ff.
Fe-hydroxides 37 f., 40
Fe-Mg-chlorites 124, 126 f.
Fe-sepiolite, a-value 84
—, DTA curve 80
—, PA-curve 81
Fe-sulfides, DTA curves 26
feldspars 88
ferromagnetic minerals 119 f.
ferromagnetism, cassiterites 119
final temperature 4, 20
fireclay-mineral 91
—, determination of the degree of disorder 131 f.
—, DTA curve 65
—, DTA data 67
—, PA-curve 66
flint 147, 152 f.
—, DTA data 150 f.
fluorides 31 f.
fluorite 31 f.
—, DTA curve 32
—, DTA data 32
form, decomposition peak, criterion for structural disorder 107
framework silicates 88
freezing 3
furnace atmosphere 5, 9 f., 12

galena 28
—, DTA curve 27
—, DTA data 30
garnets 63
gas density, combined with DTA 23
gaseous reaction products 9, 22
—, decomposition of chalcogenides 25
gaylussite 50, 53
—, DTA curve and data 51
—, PA-curve 52 f.
germanite 28
—, DTA curve 26
—, DTA data 29
gibbsite, DTA curve 38
—, DTA data 39 f.
—, PA-curve 40
gismondite, DTA curve 88
—, DTA data 90
—, PA-curve 89
glauconite 64, 73, 94
—, DTA curve and data 74
goethite 38, 84 ff.
—, dehydration, heat of reaction 20
—, DTA curve 38
—, DTA data 39
—, PA-curve 40
—, peak temperature lowering by Al-incorporation 114 f.
goslarite 35
—, DTA curve 34
—, DTA data 36
grain size, influence on DTA 8, 10, 12 64, 73, 114 f.
— —, chlorites 123
— — distribution, kaolinites 132
— —, DTA estimation in hydrated clay minerals 91, 130
— —, kaolinites 131 f.
graphite 24 f.
—, distinction from sulfur, sulfides and organic matter 25
—, DTA curve 25
grimselite 55
grinding, sheet silicates 64, 79
—, standard mineral treatment 12
grochauite 126
gypsum 6, 34 f.
—, DTA curve 34
—, DTA data 36
—, PA-curve 35
—, sensibility of proof 17

Subject Index

haidingerite, DTA curve 59
—, DTA data 58
—, PA-curve 62
halite 31
—, DTA curve and data 32
halloysite 65, 91, 94
—, DTA curve 65
—, DTA data 67
—, grain size estimation by DTA 130
halogenides 31 ff.
—, DTA curves and data 32
hausmannite 37
—, DTA data 37
heat capacity 6, 7
— change 2, 5 ff.
— conductivity 6 ff.
heat of reaction 14 ff., 18, 36 f.
— —, determination 20 f.
— —, —, examples: quartz, goethite, clay minerals, carbonates 20
— —, kaolinite 64
heating-cooling hysteresis, cristobalite 137, 139
heating rate 5, 7, 10 ff., 18, 20, 93
— —, Fe-carbonates 41
— —, variation, determination of thermodynamic equilibrium temperatures 19
heazlewoodite 25
hectorite 71, 128 f.
—, DTA curve 69
—, DTA data 72
—, PA-curve 70
hematite 40, 116, 119
—, coloured jasper 152
—, DTA data 37
hemimorphite 63
—, DTA curve 63
heulandite, DTA curve 88
—, DTA data 90
—, PA-curve 89
high-low inversion, quartz 10 f., 15, 17, 36 f., 140 ff.
—, sulfides 16
high-pressure DTA 21 ff.
high-quartz 140 f.
high-temperature X-ray, combined with DTA 21
hisingerite 91
—, DTA data 92
homogenization, heating of magnetite – ulvite intergrowths 121

hornblende, DTA curve 63
hornstone 147, 152 f.
—, DTA data 150
huntite 41
—, DTA curve 43
—, DTA data 44
—, PA-curve 45 f.
hydrated carbonates 50 ff.
— —, decomposition 55 f.
— sulfates 33 f., 36
— —, preparation 35
hydration behaviour, smectites and vermiculites 71
hydrogrossular 63
hydromagnesite 41, 53
—, DTA curve and data 54
—, PA-curve 55
hydromicas 72 f.
hydrotalcite 53
—, DTA curve and data 54
—, PA-curve 55
hydroxides 37 ff., 94, 107, 125
—, DTA criteria for disordered structure 38
hydrozincite 114
—, DTA curve and data 49
—, PA-curve 50, 52
hysteresis, cristobalite inversion 137, 139
—, dependence on synthesis temperature 138

ICTA recommendations for DTA data publication 10 f.
identification diagrams 98 ff.
— —, development 93 f.
— —, ΔT-measurement 93 f.
igneous rocks, quartz t_i 141, 143
illite 64, 67, 73, 84 ff., 94
—, dehydration, heat of reaction 20
—, DTA curves and data 74
— montmorillonite mixed layers 82
—, —, DTA curve and data 83
—, PA-curve 68
ilmenite 121
ilvaite 63
inclusions in quartz, reason for t_i-variation 142
index of disorder, kaolinites 133
indirect DTA characteristics 13, 45, 83 f.
— — — —, a-value 83

indirect DTA characteristics, PA-curve 13
— — —, standard decomposition (dehydration) temperature 13
inert substance 2, 7, 24, 27
initial temperature 4, 19f., 107
intergrowths, magnetite-ulvite 121
interlayer cation, smectites and vermiculites 129
intermediate reaction products, sulfates 33
internal standard 16, 17, 31, 57, 116, 140, 146, 148f., 153
introductory diagrams 93ff.
inverse quartz vein, t_i 154, 158
inversion, low-high cristobalite 92
— behaviour of crystals, influence of disorder 134ff.
— temperatures 10f., 15ff., 28, 30, 37, 133ff.
— —, Cu-Ag-sulfides 116ff.
— —, —, dependence on Fe-content 118
iowaite 32
iron ores 40
— —, DTA data 41
irregular mixed layers 82
— — —, a-values 84
— — —, DTA curves 83
— — —, DTA data 82
irregularity, sheet silicates 83f.
isothermal DTA 23

jalpaite 116
jasper 134ff., 138, 147f., 152
—, DTA data 150

kaemmererite 75, 123f., 126
—, DTA curve 75
—, DTA data 76
kaolinite 40, 64ff., 78, 84ff., 91, 128
—, dehydration, heat of reaction 20
—, —, pressure dependence 22
—, determination of the degree of disorder 130ff.
—, —, procedure 133
—, —, $t/\Delta T$-diagram 131
—, DTA curves 65
—, DTA data 67
—, exothermic effect 65
—, PA-curves 66
—, semi-quantitative determination 65f.

kascholong 135
kernite, DTA curve 59
—, DTA data 58
key diagrams, $t/\Delta T$ 94ff.
kieserite 33, 36
kinetic data 21, 36
— —, kaolinite 64
kotschubeite 77, 123f., 126
—, DTA data 76
kutnahorite 42, 114

laterite 39f.
—, DTA data 41
laumontite 88f.
—, DTA curve 88
—, DTA data 90
—, PA-curve 89
lawsonite 63
lepidocrocite 38, 40
—, DTA data 39
lepidolite 73
leuchtenbergite 126
—, DTA curve 75
—, DTA data 76
limonite 18, 38
—, DTA data 39
—, PA-curve 40
lizardite 78
loeweite 34
loughlinite 53
low-quartz 140ff.
low-temperature cristobalite 134ff.
— —, DTA curves 136
— —, DTA data 135
low-tridymite 134

macallisterite 57
maghemite 119, 122
—, DTA data 37
magnesite, DTA curve 43
—, DTA data 44
—, PA-curve 45
magnetic transformation 3, 36, 118ff.
magnetites, Curie temperatures 37
—, dependence on chemical composition 120f.
malachite, DTA curve 49
—, DTA data 49
manasseite 41
marcasite, DTA curve 26
—, DTA data 29
mass spectrometric DTA 22
mcinstruyite 116

Meerschaum 79f.
melanterite 35
—, DTA curve 34
—, DTA data 36
melting 3
—, borates 57
—, sulfates 33
— temperatures, chlorides 31
— —, cuspidine 33
— —, metals 14, 20, 24
— —, —, determination under high pressure 23
— —, nitrates 57
— —, phosgenite 49f.
metals, DTA determination 24
metamorphic rocks, t_i-variation 141, 144
metamorphism, cause of disorder 143f.
Mg-calcite 108
—, decomposition temperatures 108
Mg-chlorite 78, 124, 126
—, a-values 84
—, DTA curve 75
—, DTA data 76
—, PA-curve 77
Mg-Fe-chlorites 124
Mg-incorporation into calcites, DTA determination 108f.
Mg-sulfates, dehydration 33
mica 64, 71ff.
—, dehydration 71, 73
—, preparation 71
mica-montmorillonite mixed layer 81
microcrystalline quartz, inversion behaviour 134, 138, 146ff.
—, DTA curves 148f., 155
—, t_i-dependence on particle size 147
migmatites, quartz t_i-variation 145
minette 76
mirabilite, DTA curve 34
—, DTA data 36
mixed layers 64, 73, 81ff.
—, comparison with DTA data of their components 83
—, dehydration 82f.
—, DTA curves 83
—, DTA data 82
mixtures of minerals 84ff., 94
—, DTA curves 85
—, DTA data 86f.
Mn-calcite 42
—, DTA data 44

Mn-incorporation into Fe-dolomites 111
molybdenite, DTA curve 27
—, DTA data 30
montmorillonite 40, 64, 69ff., 73, 84ff., 94, 128ff.
—, a-value 84
—, dehydration, heat of reaction 20
—, DTA curves 69
—, DTA data 72
—, DTA differentiation from vermiculites 71
—, PA-curves 70
montmorines 69ff.
mounanaite 62
muscovite 64, 72
—, DTA curve and data 74

nacrite 131
nahcolite, DTA curve 42
—, DTA data 44
—, PA-curve 52f.
natrolite 88f., 91
—, DTA curve 88
—, DTA data 90
Néel temperatures 119
nepheline 89
nesquehonite 50
—, DTA curve 51
—, DTA data 51
—, PA-curve 50, 53
nickel block sample holder 6, 25, 35, 47, 85, 109, 112, 119, 123
nitrates 57
nitre 57
nontronite 71, 94, 129
—, a-value 84
—, DTA curve 69
—, DTA data 72
—, PA-curve 70
norsethite 9, 115
—, DTA curve 43
—, DTA data 45
—, PA-curve 47

olivines, Néel temperatures 119
opal 18, 92, 134
—, DTA data 92
opal-cristobalite 134
organic matter 17, 25, 92
ortho-chlorites 73

ortho-silicates 63
otayite 129
—, DTA curve 69
—, DTA data 72
overlapping effects, peak area measurement 18f.
— —, separation 50
oxidation 3, 33
—, chalcogenides 27f., 31
—, Fe-carbonates 41
—, S and sulfides 24, 27ff.
oxides 36f., 94
—, DTA data 37

packing density 8, 10, 12, 148
PA-curve 9, 43, 45, 57
—, definition 13
palygorskite 79, 94
—, a-value 84
—, DTA curve and data 80
—, PA-curve 81
paragonite 72
parisite, DTA curve 49
—, DTA data 49
—, PA-curve 45, 50
partial pressure 6
—, influence on decomposition and dehydration temperatures 9
particle size, influence on dehydration temperature 91
—, influence on t_i of cryptocrystalline quartz 147
Pb-incorporation into aragonites 109
P_{CO_2} 9
P_{H_2O} 9
peak 4
— area 7, 13, 15, 18, 20
— —, measurement 18, 65
— —, —, kaolinites 65
— —, —, overlapping 18f.
— temperature 4, 5, 13, 16, 19
— —, dependence on chemical composition and degree of disorder 94
— —, lowering by substitutions 107ff.
pelitic rocks 85
pennine 126f.
petrologic interpretations, t_i-variation of metamorphic quartz crystals 143ff.
—, t_i-variation of microcrystalline quartz crystals 153ff.
phengite 72
phlogopite 72

phosgenite, DTA curve and data 49
—, PA-curve 50
phosphates 57
—, preparation 57, 60, 62
photocell compensator 4, 15f.
—, margins of error 16
—, suppression of the zero point 16
pickeringite 34
pirssonite 50, 52
—, DTA curve and data 51
—, PA-curve 52f.
pisolite 48
piston-cylinder apparatus 22
plagioclase 88f.
plumbocalcite 108
—, DTA data 44
—, PA-curve 46
polyhalite 34
—, DTA curve 34
—, DTA data 36
polythermic dehydration 33
pore solutions, affecting structural defects 143
preparation, micas 71
—, phosphates 57ff.
—, sheet silicates 64
—, sulfides 25, 27
preparative factors 10, 12
pressure variations, influence on exothermic effects 22
primary disorder 143, 145
prochlorite 78, 126
—, DTA curve 75
—, DTA data 76
—, PA-curve 77
proustite, DTA curve 27
—, DTA data 29
pseudothuringite 78, 126
—, DTA data 76
Pt sample holders 6
—, alloying with S 25
pyrite 25, 28
—, DTA curve 26
—, DTA data 29
pyrolusite, DTA data 37
pyrophyllite, DTA curve 68
—, DTA data 67
pyroxenes 63
—, high-pressure DTA 22
pyrrhotite 28, 119
—, DTA curve 26
—, DTA data 29

quantitative DTA 18ff.
quartz 10, 40, 85ff., 134, 139
—, authigenic in sediments, t_i 37, 141, 154
—, DTA data 37, 150f., 154
—, high-low inversion temperature 10f., 15, 17, 31, 36, 37, 150f., 154
—, —, heat of reaction 20
—, incorporation of strange ions 140
—, petrologic interpretation of t_i-variations 143ff., 153ff.
—, processes of diagenetic formation 153
—, variation and average values of t_i from igneous, metamorphic and sedimentary rocks 141f.
— veins, t_i-variations 153ff.

reaction area 4, 7, 18
— temperature range 4, 20
reconstruction, crystal structure 3, 127
—, zeolites 91
recrystallization, disordered structures 38, 143f., 146
rectorite 81
reduction 3
reference material 7, 12, 15
regularity, sheet silicates 83
rehydration, smectites 129f.
replacement phenomena 153f., 158
reproducibility of measurement 2, 15
reversible structural transformations 115
rhodochrosite 42, 45, 56
ring silicates 63
ripidolite 78, 126, 128
—, DTA curve 75
—, DTA data 76
rivadavite 57
rock crystals, t_i 142
rozenite 34
rubellite, DTA curve 63
"rumpfite", DTA curve 75
rutile 121

salts 32ff.
sample arrangement 6f., 10, 12
— holder 6, 12
— —, sulfide investigations 25, 27
sandstone matrix, quartz t_i 145
sanjuanite 62

saponite 71, 128f.
—, DTA curve 69
—, DTA data 72
sassolite (sassoline), DTA curve 38
—, DTA data 39
sauconite 71, 128
schoderite 62
schoerl, DTA curve 63
schroeckingerite 53
—, DTA curve and data 54
—, PA-curve 55
scolecite 88
—, DTA curve 88
—, DTA data 90
—, PA-curve 89
secondary defects 143
sedimentary chlorites 123ff.
— micas 73
— minerals, prepared mixtures, DTA curves 85
— —, DTA data 86f.
sediments 64, 82, 84, 92, 125, 127, 130, 139, 146, 153
selenides 25
semi-quantitative mineral determination by means of the PA-curve 13, 93ff.
sensibility of proof 17
sepiolite 79, 94
—, a-value 84
—, DTA curve and data 80
sericite, DTA data 74
serpentines 64, 78f., 125
—, a-values 84
—, comparison with the DTA data of chlorites 127
—, dependence of decomposition temperature on Fe-content 79
—, DTA curves 78
—, DTA data 79
—, preparation 79
shape, decomposition peak, criterion for structural disorder 107
—, dehydration peak of re-hydrated smectites 130
—, DTA curve 4, 13
—, inversion peak, quartz, influence of disorder 134f., 137, 139
—, —, microcrystalline quartz 147
—, OH-bearing carbonates 48, 50
—, oxidation deflections 25
sheet silicates 64, 84, 94

sheet silicates, exothermic effect 83
— —, temperature interval between decomposition and reconstruction temperature 83
siderite 41, 45, 56, 112
—, DTA curve 42
—, DTA data 44
silicates, dehydration 63
silification 153, 158
silver, melting 24
— sulfides, inversion temperatures 31
simultaneous calibration 7, 15, 20
— DTA and TG 1, 23
sintering, micas 73
—, Na-Ca-carbonate 53
—, zeolites 89
smectites 69 ff., 128 ff.
—, DTA differentiation from vermiculites 71
smithsonite, DTA curve 42
—, DTA data 44
—, PA-curve 45
soda 50, 94
—, DTA curve and data 51
—, PA-curve 53
soils 13, 18, 37 ff., 40, 64, 66, 82, 84, 91 f., 125, 127, 130, 139, 145
—, DTA data 41
—, organic substances 92
solid solutions, carbonates 114
—, DTA distinction from eutectic melting 33
—, sulfides, DTA determination of the chemical composition 28, 118
specific heat 7, 18, 23
speckstone 67
sphalerite 28
—, DTA curve 27
—, DTA data 30
spinel 91
spurrite 9
Sr-incorporation into aragonites, DTA determination 109 f.
standard decomposition (resp. dehydration) temperature 13, 44 f., 49, 51, 56
— DTA conditions, proposal 12
— peak temperature 45
— — —, carbonates 57
standardization 2, 9, 11, 42, 123, 132
—, proposed conditions 12
standards, internal 16 f.

stephanite 31
—, DTA curve 27
—, DTA data 29
stilbite 89
—, DTA curve 88
—, DTA data 90
—, PA-curve 89
strohmeyerite 116
strontianite, decomposition 9
—, DTA curve 43
—, DTA data 44
—, PA-curves 46
—, structural transformations 48
—, —, influence by substitutions 115
structural decomposition 3
— defects 134, 137, 143
— disorder, sheet silicates 64, 66, 70
— stability, influence of disorder 134
— —, influencing factors 143
— —, lowering by substitutions 108 ff.
— transformations 3, 8 f., 14, 17, 21, 23, 28, 30 f., 33, 36
— —, carbonates 47 f., 115
— —, chalcogenides, influence by chemical composition 115 ff.
— —, heats of reaction 20
— —, nitrates 57
— —, SiO_2 modifications 133 ff.
struvite, DTA curve 59
—, DTA data 58
—, PA-curve 61
sublimation 3
substitutions, carbonates, influence on the decomposition temperature 108 ff.
—, chalcopyrites, S by Se 28
—, chlorites, influence on DTA data 123
—, influence on t and ΔT 107
—, magnetites, influence on Curie temperatures 119 ff.
—, palygorskite and sepiolite 79
—, quartz 140, 142
—, vermiculites 128
sudoite 69, 73 f., 78
—, DTA data 76
sulfates 33 ff.
—, DTA curves 34
—, DTA data 36
sulfides 16, 25 ff., 94
—, conditions of analysis 28
—, DTA curves 26 f.

sulfides, DTA data 29
—, high-low inversion 16, 28, 30, 116ff.
—, oxidation 24
—, preparation 25, 27
—, structural transformations 28, 30, 116ff.
sulfur 17, 24f., 27, 94
—, DTA characteristic 24
—, DTA curve 25
surface layer disorder 146f.
swelling chlorite – chlorite mixed layer 82
sylvite 31
—, DTA curve and data 32
synthetic cristobalites, DTA curves 136
— —, DTA data 137
— —, interdependence between t_i and synthesis temperature 135f.

talc 67
—, DTA curve 68
—, DTA data 67
—, PA-curve 68
talc-chlorites 73
—, DTA curves 75
—, PA-curve 77
tarnowitzite 48
—, DTA curve 110
t/ΔT-diagrams 93ff.
tellurides 25
temperature difference, ΔT 4ff.
— —, measurement 93f.
temperature gradient, sample 8
temperature measurement, margin of error 15f.
tennantite 31
—, DTA curve 26
—, DTA data 29
teruggite 57
thenardite, DTA curve 34
—, DTA data 36
thermal gasvolumetry 23
— optics 21, 36
thermobalance 1
thermocouples 6ff., 15
—, arrangement 7, 12, 17
thermodynamic data 18f.
— equilibrium temperatures 4, 7, 13, 19, 21, 33
— — —, DTA determination 19
thermogravimetry 1, 23, 63
thermomanometry 23

thomsonite 88f.
—, DTA curve 88
—, DTA data 90
—, PA-curve 89
thuringite 73, 125f., 128
—, DTA data 76
t_i = temperature of inversion
t_i, dependence on temperature of synthesis, synthetic AlPO$_4$-phases 146
—, — —, synthetic cristobalites 137
t_i, diagenetically silicified ore vein quartz 155
Ti-incorporation into magnetite, DTA determination 120f.
t_i-isotherms 145
t_i-variation, contactmetamorphic quartz 141, 145
titanohematite 119
titanomagnetites 119ff.
tosudite 81
tourmalines 63
—, DTA curves 63
transformation temperatures 14ff.
tridymite, t_i 37, 142, 152
trona 50
—, DTA curve and data 51
—, PA-curve 52f.

ulexite 57
—, DTA curve 59
—, DTA data 58
ulvite 121

vanadates 62
Verkieselung 153, 158
vermiculites 64, 69ff., 94, 128ff.
—, a-values 84
—, dependence of thermal effects on chemical composition 71
—, DTA curves 69
—, DTA data 72
—, DTA determination of the di-octahedral part 128
—, PA-curves 70
—, substitutions 128
vesuvianite 63
vivianite, DTA curve 59
—, DTA data 58
—, PA-curve 60
voglite, DTA curve and data 54
—, PA-curve 55
volatile reaction products 9, 22

water, adsorption, allophanes 91
—, — capacity, smectites and vermiculites 71
—, interlayer ~, smectites 130
—, palygorskite and sepiolite 79
—, zeolites 88
weathering 140, 143
whewellite 92
witherite 9, 115
—, DTA curve 43
—, DTA data 45
—, PA-curve 46
wolframite solid solutions, Néel temperatures 119
wurtzite, DTA curve 27
—, DTA data 30

X-ray, combined with DTA 21

zaratite, DTA curve and data 54
—, PA-curve 46
zeolites 88 ff.
—, decomposition 89
—, dehydration 88
—, DTA curves 88
—, DTA data 90
—, PA-curves 89
—, transformations 89
zero line 4, 18
— point suppression, photocell compensator 16
zinnwaldite 72 f.
—, DTA curve and data 74

Minerals and Rocks

As from volume 10 the series formerly titled Minerals, Rocks and Inorganic Materials will be continued under the new title.
Editor-in-Chief: P.J. Wyllie
Editors: W. von Engelhardt, T. Hahn

Vol. 1: W.G. Ernst
Amphiboles
Crystal Chemistry, Phase Relations and Occurrence
With 59 figures
X, 125 pages. 1968
ISBN 3-540-04267-9
Cloth DM 30,–
ISBN 0-387-04267-9 (North America)
Cloth $11.20
Subseries: Experimental Mineralogy

Vol. 2: E. Hansen
Strain Facies
With 78 figures. 21 plates
X, 208 pages. 1971
ISBN 3-540-05204-6
Cloth DM 58,–
ISBN 0-387-05204-6 (North America)
Cloth $19.60
Distribution rights for U.K., Commonwealth and the Traditional British Market (excluding Canada): Allen & Unwin, Ltd., London

Vol. 3: B.R. Doe
Lead Isotopes
With 24 figures
IX, 137 pages. 1970
ISBN 3-540-05205-4
Cloth DM 36,–
ISBN 0-387-05205-4 (North America)
Cloth $14.80
Subseries: Isotopes in Geology

Vol. 4: O. Braitsch
Salt Deposits – Their Origin and Composition
Translated from the German edition by P.J. Burek and A.E.M. Nairn in consultation with A.G. Herrmann and R. Evans.
With 47 figures
XIV, 297 pages. 1971
ISBN 3-540-05206-2
Cloth DM 72,–
ISBN 0-387-05206-2 (North America)
Cloth $29.60

Vol. 5: G. Faure, J.L. Powell
Strontium Isotope Geology
With 51 pages
IX, 188 pages. 1972
ISBN 3-540-05784-6
Cloth DM 48,–
ISBN 0-387-05784-6 (North America)
Cloth $16.90
Subseries Isotopes in Geology

Vol. 6: F. Lippmann
Sedimentary Carbonate Minerals
With 54 figures
VI, 228 pages. 1973
ISBN 3-540-06011-1
Cloth DM 58,–
ISBN 0-387-06011-1 (North America)
Cloth $23.70

Vol. 7: A. Rittmann
Stable Mineral Assemblages of Igneous Rocks
A Method of Calculation
With contribution by V. Gottini, W. Hewers, H. Pichler, R. Stengelin
With 85 figures
XIV, 262 pages. 1973
ISBN 3-540-06030-8
Cloth DM 76,–
ISBN 0-387-06030-8 (North America)
Cloth $31.10

Vol. 8: S.K. Saxena
Thermodynamics of Rock-Forming Crystalline Solutions
With 67 figures
XII, 188 pages. 1973
ISBN 3-540-06175-4
Cloth DM 48,–
ISBN 0-387-06175-4 (North America)
Cloth $22.10

Vol. 9: J. Hoefs
Stable Isotope Geochemistry
With 37 figures
IX, 140 pages. 1973
ISBN 3-540-06176-2
Cloth DM 39,–
ISBN 0-387-06176-2 (North America)
Cloth $16.00

Vol. 10: J.T. Wasson
Meteorites
Classification and Properties
With 70 figures
X, 316 pages. 1974
ISBN 3-540-06744-2
Cloth DM 76,–
ISBN 0-387-06744-2 (North America)
Cloth $31.10

Prices are subject to change without notice

Springer-Verlag
Berlin
Heidelberg
New York

Crystal Chemistry of Non-Metallic Materials

Editor: R. Roy

Vol. 1: Roy/Newnham: **Principles of Crystal Chemistry**
In preparation

Vol. 2: Newnham: **Properties of Solids in Relation to Structure**
In preparation

Vol. 3: O. Muller, R. Roy: **Major Binary Structural Families**
In preparation

Vol. 4: O. Muller, R. Roy: **The Major Ternary Structural Families**
46 figures. IX, 487 pages. 1974
ISBN 3-540-06430-3 Cloth DM 76,–
ISBN 0-387-06430-3
(North America) Cloth $31.10
Prices are subject to change without notice

Contents: The A_2BX_4 Structures. The ABX_4 Structures. The ABX_3 Structures. Other Ternary Structure Families.

The purpose of the book is to provide an up-to-date crystal-chemical review of the major ternary structure families: the A_2BX_4, ABX_4 and ABX_3 structures, which include the majority of the phases of interest to materials scientists, ceramists, solid-state physicists, mineralogists and inorganic chemists.

A brief introductory chapter explains the scope of the book and discusses ionic radii, coordination formulae and other terms of nomenclature used.

The second, third and fourth chapters discuss the A_2BX_4, the ABX_4 and the ABX_3 structure families, respectively. Each is subdivided into three parts: (1) a brief description of the major structures within that family and their importance, (2) the largest section treating the individual structures and (3) a final discussion of the interrelationships (such as transitions) among these structures. In Chapter V a very brief summary is given of some other ternary structure families.

The book is provided with an index and a large number of references. The appendix contains exhaustive supplementary tables which are particularly useful for detailed references. There is enough factual information here to serve as a reference source at least for the next few years.

Springer-Verlag Berlin Heidelberg New York 1974